Shaping the Earth

Shaping the Earth

Tectonics of Continents and Oceans

. . .

READINGS FROM
SCIENTIFIC AMERICAN MAGAZINE

Edited by

Eldridge Moores

University of California at Davis

W. H. FREEMAN AND COMPANY
New York

Some of the SCIENTIFIC AMERICAN articles in *Shaping the Earth: Tectonics of Continents and Oceans* are available as separate Offprints. For a complete list of articles now available as Offprints, write to Product Manager, Marketing Department, W. H. Freeman and Company, 41 Madison Avenue, New York, NY 10010.

Library of Congress Cataloging-in-Publication Data

Shaping the earth : readings from Scientific American
   Magazine / edited by Eldridge Moores.
       p.   cm.
   Includes bibliographical references and index.
   ISBN 0-7167-2141-4
   1. Geodynamics.   I. Moores, Eldridge M., 1938–
II. Scientific American.
QE505.S48   1990
551—dc20   90-3585
                                              CIP

Printed in the United States of America

1 2 3 4 5 6 7 8 9 0   RRD   9 8 7 6 5 4 3 2 1 0

CONTENTS

SECTION III: MAKING MOUNTAINS

Note on cross-references to SCIENTIFIC AMERICAN articles: Articles included in this book are referred to by chapter number and title; articles not included in this book but available as Offprints are referred to by title, date of publication, and Offprint number; articles not in this book and not available as Offprints are referred to by title and date of publication.

Preface

Viewed through the prism of time, we see the earth as a dynamic entity. Earthquakes shake and rearrange topography. Volcanoes spew forth gases and dust that affect global climatic changes. Mountains rise high and are maintained for millenia by uplifting action greater than the erosive power of raging rivers. Over a span of millions of years, oceans open and close, and, as a result, continents drift apart, converge and collide.

We see evidence of these processes at work in the world today. Newspapers and news broadcasts report earthquakes, volcanic eruptions, floods and other catastrophic natural events as they happen around the globe. Yet most of the processes that shape the earth are slow in comparison to the life span of even the oldest humans. Major earthquakes in any geologically active region may occur once in a hundred or more years. Active volcanoes may be quiescent for centuries before erupting. The speediest mountains may be uplifting only a few millimeters per year. Continents move apart at the rate of a centimeter or two per year, roughly the rate of growth of a fingernail. Slow as geological processes are, their accumulation over the vast span of time produces great changes in the landscape and geography of the earth, patterns of climate, evolution and even the course of human history.

In a broad sense, geology is the study of how the earth has changed through time. Clues to the history of the earth and the geological processes that have shaped it are hidden in the rocks that form the continents and ocean floors. By searching out and discovering patterns in these clues, geologists attempt to decipher that history. Thus, geology is an extension of history and archaeology to a much longer time scale using a less well-preserved record.

In order to understand how earth processes work, geologists use a model-deductive method of inquiry. Before a model can be developed, geologists must make many careful, and often diverse, observations. The information gathered is then synthesized into an overall conceptual model (or, as it is sometimes called, a paradigm) from which one can deduce a series of corollaries. These corollaries may be in the form of consequences or predictions. If the model is valid, then its corollaries should be so as well. A good model may not necessarily be correct, but it must be testable and capable of modification.

Geologists check the truth of their models by making new observations and comparing them to the theoretical model. If the observations fit the model, the model is assumed to be true. If the observations do not fit, then the model must be adjusted to accommodate the new observations.

Models are an indispensable part of geological study. (You will find many references to conceptual models and their use in this book to describe how the earth's dynamic processes work.)

Over the past thirty years, the science of geology has been in great and exciting ferment, occasioned by the development of a major new comprehensive model of the geologic history of the earth. The model is called "plate tectonic theory." (*Tectonic* refers to the history of the formation and deformation of the major regional global architectural features of the earth, such as mountain chains, ocean basins, continents and so on.)

Plate tectonic theory was a remarkable breakthrough in our understanding of the solid earth. It effectively unified all previously disparate branches of earth science into a single, simple world view. The theory, in brief, holds that the outer part of the solid earth consists of a relatively cool, strong, dense shell, called the lithosphere, riding on a hotter, weaker, less dense zone, called the asthenosphere. As defined, the lithosphere includes both the outermost layer of the earth, the crust, and the upper part of the layer below, the mantle, down to a combined depth of about 100 kilometers. The asthenosphere extends from about 100 to 300 kilometers in depth.

The lithosphere is divided into a number of segments called plates, which are in motion with respect to each other (as shown in the figure). Thus, plate tectonic theory focuses on the formation and deformation of the plates as they move during the course of earth's history.

The plates are somewhat rigid bodies that interact with each other along their boundaries. Three kinds of boundaries exist. Divergent boundaries or mid-ocean ridges are boundaries where plates spread apart and new plates are created by upwelling of material from the earth's interior. Convergent or subduction zone boundaries are those where one plate dives beneath another and is recycled back down inside the earth. Conservative or transform fault boundaries are those where one plate slides past another, and lithosphere is neither created nor destroyed. Places where three plates meet and interact are known as triple junctions.

Most of the present dynamic activity of the earth is concentrated at the plate boundaries. Almost all the earthquakes that occur on earth are confined to narrow zones along the plate boundaries. In addition, almost all volcanic activity occurs in response to divergent or convergent plate margin activity.

Continents are passive riders on the plates. Be-

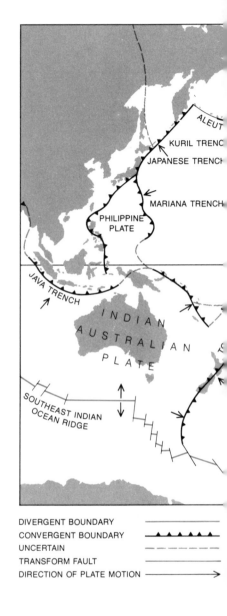

DIVERGENT BOUNDARY	———————
CONVERGENT BOUNDARY	—▲▲▲▲▲—
UNCERTAIN	– – – – –
TRANSFORM FAULT	———————
DIRECTION OF PLATE MOTION	——————→

cause continental crust is thicker than oceanic crust, continents tend to stick up above the surface of the ocean. If a divergent plate margin forms in the middle of a continent, it splits apart, and the two parts move away from each other as an ocean, such as the Atlantic Ocean, forms in the split. In some places, such as along the western side of South America, oceanic plates descend beneath the margin of a continent, producing volcanic-rich mountains such as the Andes. In other regions, a continent riding on a

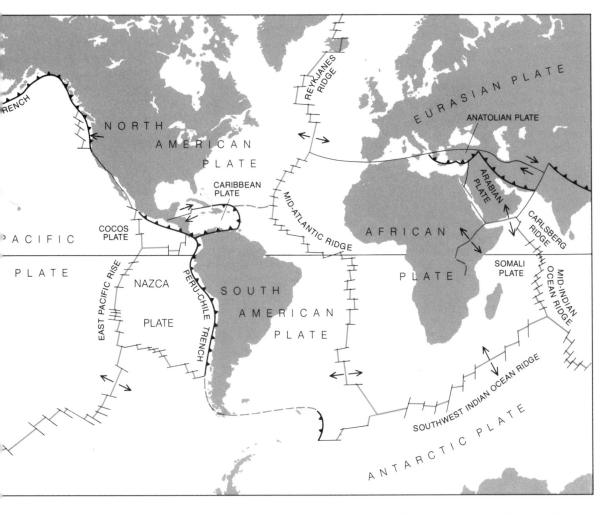

**MAJOR LITHOSPHERIC PLATES and their boundaries
are mapped. The lithosphere consists of the crust and the
rigid upper part of the mantle. In midocean, away from the
spreading center, the lithosphere is about 100 kilometers
thick. The upper five to seven kilometers is crust. The
lithospheric plates are in motion with respect to one another, and the boundaries are defined according to this
relative motion. At divergent boundaries the plates move
apart; the spreading centers at the midocean ridges are
divergent boundaries. At transform faults the plates slide
past each other. At convergent boundaries the plates move
together and one plate plunges under the other in the
process of subduction. Thus oceanic crust is created at divergent boundaries and destroyed at convergent boundaries.**

plate being subducted converges and ultimately collides with another, thereby forming great mountain
ranges. The Alpine-Himalayan mountain system,
which stretches from western Europe to eastern
Asia formed as Africa, Arabia and India approached
and collided with Eurasia.

Geology is a relatively young science. It originated only about two hundred years ago with studies by European and North American naturalists

interested in their surroundings. Until the end of
World War II, the rocks of the continents were the
principal focus of geologic study, and a considerable
body of knowledge and theory had been accumulated on the origin of geologic features of the continents. Most people thought that the continents were
fixed and that the ocean basins were very old, but
very little was known about the geology of the
ocean basins.

In the 1950's and early 1960's a great deal of research was done on the geology buried beneath the depths of the oceans using geologic and geophysical techniques aboard oceanographic ships. As knowledge of the oceans accumulated, it became increasingly clear that the existing explanations for continental features not only did not explain the new knowledge about the oceans, but it also was not a satisfactory explanation for the geologic features of the continents. New evidence indicated that the continents were not fixed and had moved around on the surface of the earth. Data collected from the ocean basins indicated that they were not very old, but in fact were quite young. Thus, a crisis developed between the old tectonic ideas and the accumulating new knowledge. Formulation of the plate tectonic model provided a new view of the earth that resolved this crisis between observation and theory.

The formulation of the plate tectonic model probably represents the most important development in geology in the past century. The great power of the plate tectonic model is that it accounts for much of the tectonic history of the earth, as revealed in the continents and in the ocean basins. Most workers now believe that plate tectonics has characterized earth tectonics for at least the last half of its history, and perhaps more.

Despite the great simplicity of the plate tectonic model, observations made since its formulation point up the complications of the real world as compared with the model. Plate boundaries are not simple lines. They have a finite width, usually tens or even hundreds of kilometers wide. Plates are not completely rigid, as the theory originally held. They undergo deformation by folding and faulting at their edges, and in some cases they may deform in their interiors. The relationship of plate motion to the internal structure of the earth is not yet understood, although recent work provides some tantalizing clues.

The plate tectonic revolution continues to offer promising opportunities for discovery as geologists relook at the world from this new perspective. The continuing development of new investigative tools augments their ability to study the internal structure of the earth. The chapters in this book provide an introduction to the current state of knowledge of plate tectonics, the use of some new tools and some topics for future fieldwork and study.

Eldridge Moores

I

THE DYNAMIC EARTH

. . .

Introduction

This first section presents an overview of the nature of the continents and oceans, of their current tectonic activity and of the record they provide of past tectonic movements. It also discusses the possible nature of the earth's mantle beneath the tectonic plates and how processes within it may relate to plate movements.

Continental crust forms approximately 45 percent of the surface of the earth. It consists of rocks as old as 4 billion years of two general types: areas of little deformed horizontal layers of sedimentary and volcanic rock, and linear belts of folded and faulted sediments, metamorphic rocks and igneous rocks. The latter regions record the history of plate tectonic processes going back at least 2.5 billion years ago and perhaps earlier. Chapter 1, "The Continental Crust," outlines the current state of knowledge of the continents.

Perhaps the greatest discoveries of the past two centuries of geologic study is the immensity of geologic time. We now know that the Earth and the rest of the solar system is approximately 4.6 billion years old (or 4.6×10^9 years). To get an idea of the immensity of this number, think of each year as being one millimeter. All of recorded history (since 3000 B.C.) represents five meters, or roughly 16.5 feet. The earliest hominids are about 4 million years old, or 4 kilometers worth of millimeters. The last dinosaurs became extinct 65 million years ago, equivalent to 65 kilometers or 40 miles worth of millimeters. The oldest shelly fossils are 570 million years old, or 570 kilometers (roughly the distance from San Francisco to Los Angeles, from Boston to Washington, D.C., from Vancouver to Calgary or from Sydney to Melbourne). Therefore, the age of the solar system, including the Earth, in millimeters is 4,600 kilometers (2,900 miles), or roughly the distance from London to Teheran or from San Francisco to Boston.

The plate tectonic model originally was conceived as a means to explain the present activity and structure observed on the earth. A consequence of the theory was that the oceans are relatively young. We now know that the present ocean crust of the earth is less than 200 million years old (less than 200 kilometers in our one-millimeter-equals-one-year analogy). Thus, the present oceans preserve a record of less than the past 5 percent of earth history. Any record of the first 95 percent of earth history is limited to the continents.

Although the oceans are young, they are where most of the current "tectonic action" is. Oceans contain the midocean ridges as well as most of the transform faults and subduction zones. Therefore we must understand the tectonic features of the oceans if we are to gain an understanding of the tectonic processes of the earth as a whole. Because the oceanic crust is thinner than continental crust and most of it is covered by water, it is necessary to use sophisticated geophysical techniques to study it. Chapter 2, "The Oceanic Crust," describes several of these modern techniques in the course of its discussion of the nature of the ocean crust and the tectonics of oceanic plate boundaries.

Chapter 3, "The Earth's Hot Spots," describes what are thought to be the product of upwelling plumes of hot mantle material that rise into the asthenosphere from deep within the mantle, perhaps even from the core-mantle boundary. Some of this hot material burns its way through the lithosphere and produces volcanism at the surface, usually away from a plate boundary in a midplate location. As the plate moves over a hot spot, the latter leaves a trail, often of volcanoes, such as in the Hawaiian Islands. These features represent an indication of the motion of the plate relative to the hot spot. Study of these hot-spot "tracks" refines our understanding of the paths that plates follow during their movement.

The lithospheric plate motions and the deep mantle plumes are all part of a complex convection system involving the transfer of material from deep within the earth to the surface and back again. The possible patterns of convective transfer has been the subject of many theoretical studies, but the difficulty of direct observation of the pattern in the earth has made it impossible to evaluate effectively the various models of mantle convection.

Chapter 4, "Seismic Tomography," presents results from application of computer-assisted tomography CAT scanning techniques to the mantle. The technique is similar to that used in medicine, but uses seismic waves instead of Xrays. It identifies regions of seismic velocity that are slow and fast relative to the standard model of a simple radial variation in seismic velocity. The slow and fast areas are thought to be hot and cold, respectively. The results show that the pattern of convection is considerably more complex than many of the theoretical models. The effects of midoceanic ridges are reflected in slow velocities at 150 kilometers beneath them. At deeper levels, however, the pattern becomes segmented and displaced significantly from a position directly beneath the surface expression of the ridges.

One outstanding problem of plate tectonics and mantle convection has been whether or not plates descend deeper than the 700 kilometers maximum depth recorded for deep earthquakes. If lithospheric plates descend no deeper than 700 kilometers, then they must form part of a shallow convection system in the outer parts of the mantle. Recent seismic tomographic analyses suggest that high-velocity slabs are present beneath subduction zones to depths of at least 2,000 kilometers, which also suggests that in some places lithospheric plates may descend even to the core-mantle boundary.

The Continental Crust

It is much older than oceanic crust, some of it dating back nearly four billion years. It is constantly reworked, however, by cycles of tectonics, volcanism, erosion and sedimentation.

• • •

B. Clark Burchfiel
September, 1983

For a little more than 200 years earth scientists have been studying the geology of the continents, hoping to reconstruct a record of the history of the earth. It is a daunting project. The crust underlying the oceans is rapidly created; it remains intact and relatively undeformed for most of its short lifetime, and then it is rapidly destroyed. The oldest crust in the earth's ocean basins today is less than 200 million years old. In contrast, the crust making up the continents is created and modified by a variety of physical and chemical processes, often undergoing many phases of deformation and reworking that produce a complex worldwide pattern in which belts of deformed rock hundreds of kilometers wide and thousands of kilometers long are invaded by intrusions of igneous rock and locally overlain by a thin veneer of younger sedimentary rock. In addition continental crust largely resists the processes that destroy oceanic crust. The most ancient parts of the continents are about 3.8 billion years old. Thus the continental crust holds a complex and fragmentary record of the evolutionary and dynamic processes operating through 85 percent of the earth's 4.6-billion-year history (see Figure 1.1).

Continental crust underlies the continents and their margins, and also small shallow regions in the oceans. In total it covers about 45 percent of the earth's surface and makes up about .3 percent of the mass. It is distinguished from oceanic crust and from the mantle under it by its physical properties and chemical composition. The horizontal boundaries between continental crust and oceanic crust are poorly defined; they are under not only the ocean's water column but also, in most places, a thick sequence of sedimentary rock. Seismic, magnetic and gravitational data indicate that the boundary is less than 10 kilometers across in some places but is several tens of kilometers across in others. Studies of the rock of oceanic and continental crust and the correlation of rock compositions and seismic velocities indicate, however, that the oceanic crust is characterized by the igneous rock basalt whereas the continental crust is an assemblage of igneous, metamorphic and sedimentary rocks enriched in elements such as potassium, uranium, thorium and silicon.

The vertical boundary between the mantle and the crust (both oceanic and continental) is called the Mohorovičić discontinuity, more commonly the Moho. It is a zone, less than one kilometer thick in some places but several kilometers thick in others,

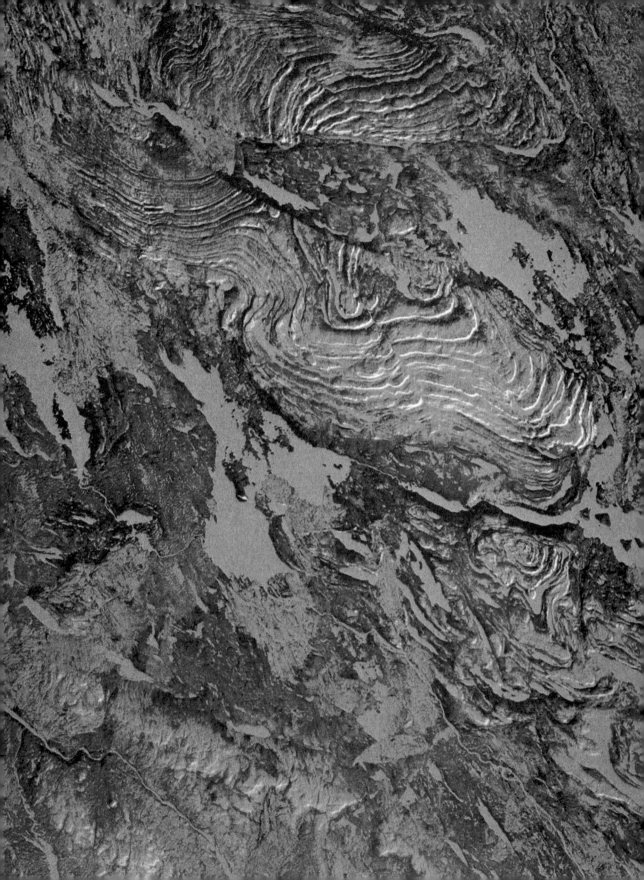

where the velocity of compressional seismic waves increases from about 6.8 kilometers per second in the crust to 8.1 in the mantle. The change in seismic velocity is caused largely by a change in the composition of the medium. Rocks of the mantle differ from rocks of the crust: they are poorer in silicon but richer in iron and magnesium.

Seismic studies of the Mohorovičić discontinuity indicate that the oceanic crust is typically from five to eight kilometers thick, whereas the continental crust varies from about 10 kilometers to more than 70. To a first approximation the crust behaves as if it were floating on the mantle. Oceanic crust is relatively thin and dense (3.0 to 3.1 grams per cubic centimeter); thus the parts of the earth's surface underlain by oceanic crust are usually far below sea level, at a depth ranging from 2,500 to 6,500 meters. Continental crust is thicker and is notably less dense (2.7 to 2.8 grams per cubic centimeter); thus the parts of the surface formed by continental crust lie near sea level or above it. The thickest parts of the continental crust usually underlie places of great elevation, such as the Himalayan and Andean mountain chains. Conversely, the thinnest parts of the continental crust usually lie below sea level at places such as the continental margins of the Atlantic.

Some important exceptions to this pattern are found at midocean ridges and some areas on the continents where volcanism is active and the crust is being stretched. In such places hot material from the deeper mantle rises to shallower levels, making the upper mantle hotter and so less dense than normal. The resulting buoyancy raises the surface elevation. The Basin and Range province of the western U.S. is a good example. The crust there is thin, but the surface elevation is nonetheless high.

Other exceptions to the pattern are found in areas with great topographic relief, where the crust bends downward over short horizontal distances, usually about 200 kilometers. Evidently the crust and the uppermost mantle deform like an elastic sheet to support the topographic load. One result is that long, linear troughs filled with sediment as much as eight kilometers deep form next to many great mountain chains. The troughs are underlain by crust of normal thickness. The crust has simply bent downward to support the weight of the mountains.

In general the rocks that form the continental crust fall into two major groups: widespread, relatively undeformed accumulations of sedimentary or volcanic rocks on the one hand and long, deformed belts of sedimentary, igenous and metamorphic rocks on the other (see Figure 1.2). The belts are called orogenic belts, after the Greek *oros*, mountain. The first group is not ubiquitous on the continents, but where it is present it always overlies the second. In places such as the central U.S. it forms a thin veneer no more than a few kilometers thick. Elsewhere, along continental margins and in linear, circular or irregular depressions in the continents, it forms sequences of rocks that can be more than 10 kilometers thick.

The second group makes up the bulk of the continental crust. Its great variety of rock assemblages gives it a heterogeneity contrasting significantly with the relative homogeneity of the oceanic crust. Each belt in the second group evolved over a span of time as long as several hundred million years, and the ages of adjacent belts may differ by hundreds of millions of years, or even a billion years; hence each belt represents a different segment of earth history. Often the younger belts are oblique to the older ones, so that the younger truncate the older. In other places belts are parallel.

A detailed examination of the rocks in the belts shows that many older belts are similar to the ones formed in the more recent geologic past. They also resemble the belts that the earth's tectonic activity is forming today. This offers the prospect that the study of modern rock formations and how they are currently deforming will yield understanding of the processes responsible for the formation of the older orogenic belts making up most of the continental crust. The theory of plate tectonics is crucial in such a venture, because the theory provides a framework in which rock assemblages and their deformation can be related to interactions of the plates that make up the entire crust. To be sure, the theory was developed largely from data gathered in the

Figure 1.1 LABRADOR FOLD BELT, shown in a Landsat image made above north Quebec, exemplifies the evolution of the continental crust; the belt was once a chain of mountains raised by the collision of two continents 1.8 billion years ago. Since then the mountains have eroded, exposing the deeper, mostly metamorphic and igneous rocks that the collision had deformed into myriad folds. Moreover, subsequent plate collisions have reshaped the continents. Nevertheless, the belt resembles the ones being formed by tectonic activity along continental margins today.

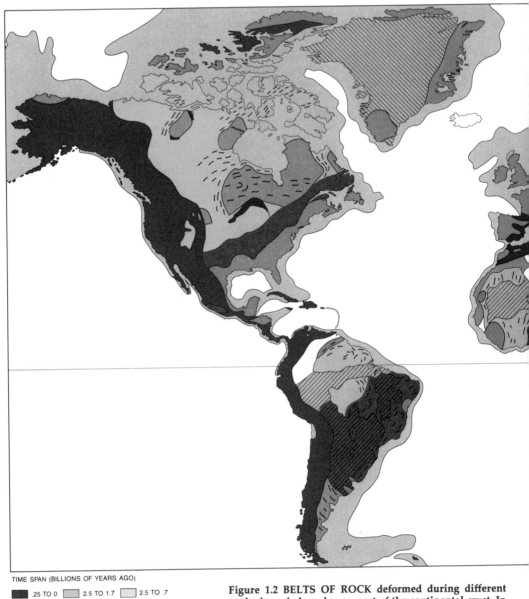

TIME SPAN (BILLIONS OF YEARS AGO)

.25 TO 0	2.5 TO 1.7	2.5 TO .7
.7 TO .25	3.8 TO 2.5	3.8 TO 1.7
1.7 TO .7		

Figure 1.2 BELTS OF ROCK deformed during different geologic periods make up most of the continental crust. In places the rocks are present beneath sedimentary or volcanic rock or ice (*black hatching*). Colors show the time

oceans, and its application to the study of continental crust has met with mixed results. Still, a modified version of the plate-tectonic concepts is a basis for understanding continental development.

The central concept of plate tectonics is straightforward: the outermost shell of the earth, the litho-sphere, can be divided into six major plates and many smaller ones that move with respect to one another with velocities ranging from a few centimeters per year up to 20 or more. The plates consist of oceanic and continental crust together with some of the underlying mantle; the Mohorovičić discontinu-

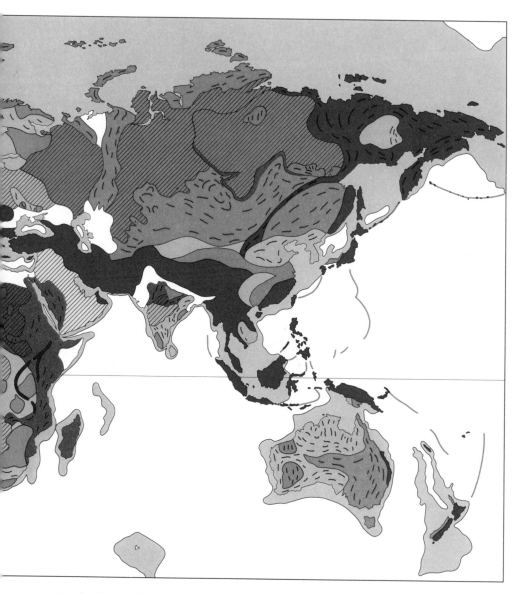

spans of major deformation episodes. Also shown are continental rocks beneath the oceans in continental margins and oceanic plateaus (*light blue*), island arcs built on oceanic crust over subduction zones (*orange*) and rift regions where the continents have been or are being pulled apart (*gray*).

ity lies within them. The plates are generally regarded as rigid bodies, so that most of their interactions are concentrated along plate boundaries, which can be zones of intense deformation. The boundaries can be classified into three basic types: divergent, transform and convergent. At divergent boundaries new oceanic crust is created; at trans-form boundaries it slides past the crust of a neighboring plate; at convergent boundaries it plunges into the mantle. The continental crust generally resists this subduction, largely because it "floats" on the mantle (see Figure 1.3).

This basic scheme must be modified in several important ways if it is to help illuminate the evolu-

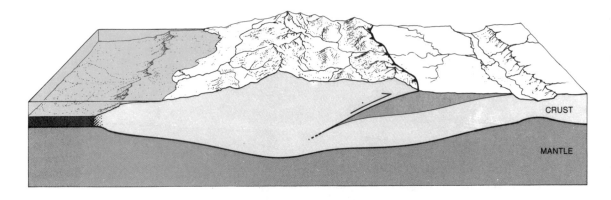

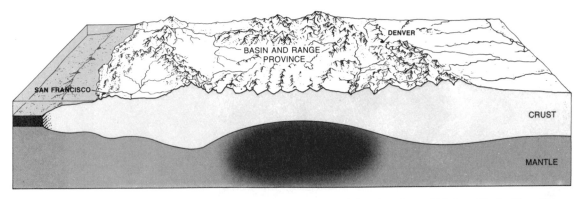

Figure 1.3 FLOTATION EQUILIBRIUM describes the relation of the continental crust to the underlying mantle. The crust, which is the lighter of the two, behaves as if it were floating; thus regions of great elevation, such as mountain chains, tend to be places where the crust is notably thick. Here two exceptions are shown. In the top drawing the crust near a mountain chain has flexed downward as if it were an elastic sheet supporting a weight. In the downward-flexed region, which has filled with sedimentary rock (*color*), the crust is thicker than one would judge from the elevation of the surface. In the bottom drawing, which shows the western U.S., a part of the mantle is hotter and so less dense than usual (*color*). Its buoyancy supports the crust, which is thinner than one would judge from the elevation of the surface.

tion of continental crust participating in plate interactions often does not behave rigidly. In the upper lithosphere, therefore, the motions of plates may be absorbed partially or entirely by deformation within the continental crust. Much of the deformation can be attributed to forces arising at plate boundaries; hence the boundaries, which generally are narrow and well defin in oceanic crust, become broad and diffuse on the continents. The boundary between the Eurasian plate and the Indian plate, for example, is more than 2,000 kilometers wide in some of the places where the continental crust of one plate is in contact with the continental crust of the other.

Second, the continental crust is markedly nonuniform in its mechanical properties, consisting as it does of belts of older rock and other preexisting structures that can localize new episodes of deformation. A zone of deformation extending well into a continent may thus form structures so greatly influenced by the anisotropies in the crust that the structures are hard to relate to the tectonic activity at the border of the plate. In such a zone of deformation it is difficult to define a plate boundary: the entire zone functions as the boundary. Commonly the older rock assemblages and structures of the continents have been subjected over time to the activity of many different plate-boundary systems. Therefore the record in the rock is often fragmentary and difficult to interpret. Under these circumstances the study of modern plate-boundary systems can show how modern and ancient plate boundaries evolve. Some examples from the three types of plate boundaries can serve as an introduction to the more complex patterns that result from the superposition of several tectonic episodes.

The divergence of two plates along a divergent plate boundary that crosses the continental lithosphere begins as the crust and its underlying lithospheric mantle become stretched and attenuated (see Figure 1.4). Crustal faults develop in long, narrow zones, and within these zones the faulted crustal rocks subside differentially, forming great tilted blocks. Since the upper part of the mantle participates in the stretching, material from lower in the mantle (the hotter, more ductile level called the asthenosphere) rises to take its place, increasing the heat flow through the lithosphere. The result is the partial melting of the mantle and a characteristic volcanism of basaltic rock that is often alkalic (that is, rich in sodium and potassium).

Sometimes the divergence ends after only a few tens of kilometers of stretching, so that the zones of attenuated continental crust remain rifted scars in the continents. Some young, still active examples are the Rhine valley and associated rifts in central Europe, the East African rift valleys and the Rio Grande rift of the American Southwest. More ancient examples are the Oslo rift in southern Scandinavia, which is some 280 million years old, the Keweenawan rift of the central U.S., which is a billion years old, and the Athapuscow and Bathurst rifts in northwestern Canada, which are more than two billion years old.

In other instances the divergence continues. The attenuation of the crust thus becomes more extreme.

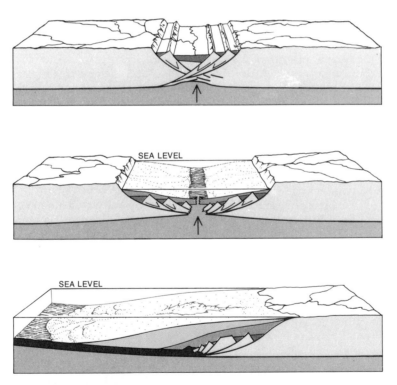

Figure 1.4 DIVERGENT PLATE BOUNDARY, where two plates move apart, causes extension and thinning of the continental crust. Initially (*top drawing*) the upper crust develops a series of brittle faults. Often the fault blocks rotate as they subside. The extension at deeper levels is less well understood. The subsidence (*middle drawing*) provides a site for the accumulation of non-marine or shal- low-marine sedimentary rocks (*medium colors*) and the rise of igneous rocks (*dark gray*). If the plates continue to diverge (*bottom drawings*), two continents result, and oceanic crust forms between them. The thinned margin of each continent subsides as it moves away from the zone of divergence; thus it is covered by unfaulted sedimentary rocks (*light color*).

It culminates in the separation of two bodies of continental crust and the formation of a new ocean basin, underlain by a widening span of oceanic crust. Each margin of the continental crust moves away from the region of divergence. The hot upper mantle moves with it. With time this hot upper mantle cools and contracts, causing the crust to subside. Meanwhile the faults that accompanied the stretching and attenuation of the crust become inactive. Sedimentary rocks begin to accumulate thick buildups above the thinned, subsiding crust. They also accumulate above the transition from continental crust to oceanic crust, forming the characteristic continental margin (called a passive continental margin) that flanks many ocean basins.

A profile through a rifted margin illustrates its evolutionary history. Thinned and faulted continental crust is overlain by a sequence of sedimentary and volcanic rocks deposited in faulted rifts during the initial phase of divergence. Those rocks are overlain in turn by a thick blanket of unfaulted sedimentary rock deposited during the latter, more gradual phase of subsidence. Well-studied examples of rifted margins are the Atlantic margins of the East Coast of the U.S. and the west coast of Africa. The margins generally have high temperature gradients during their early evolution, and so they are favorable sites for the maturation of organic matter into deposits of petroleum and natural gas. Occasionally the divergence of two continental bodies occurs near an older continental margin, and fragments of continent are rafted away to form small plateaus of continental crust partially or entirely submerged in the oceans and surrounded by oceanic crust. Examples include the Lord Howe Rise (with its highest part, New Zealand) in the southwestern Pacific and part of the Kerguelen and Mascarene plateaus in the Indian Ocean.

At transform boundaries, where two plates slide horizontally past each other along vertical or nearly vertical faults, crust is neither created nor destroyed. The horizontal displacement along the length of the boundary can measure hundreds of kilometers, even thousands. When the boundary crosses continental crust, the displacement is commonly distributed across a zone of faults as much as several hundred kilometers wide. Preexisting belts are shifted laterally, and parts of them can be rotated, greatly disrupting their original continuity. Moreover, offsets or bends in the faults can give rise to local regions of divergence or convergence.

Here are two examples. The Alpine fault in New Zealand is part of a transform system along the boundary between the Pacific plate and the Indian plate. It passes through a fragment of continental crust that was rafted away from Australia about 100 million years ago. The horizontal displacement along the fault is now about 400 kilometers, but the movement of the plates is not limited to displacement. In addition rock assemblages and structures created by activity at older plate-boundary systems have been rotated and bent, recording a total of about 1,200 kilometers of differential motion. It can be shown that the motion was purely transform about 40 million years ago and later became oblique, with components of both transform and compression. The compression has thickened the crust and raised a chain of high mountains: the New Zealand Alps.

The Dead Sea fault zone in the Middle East is a transform system connecting a divergent plate boundary in the Red Sea to a convergent plate boundary in the Taurus Mountains of southern Turkey (see Figure 1.5). In places the fault zone steps to the west, cutting across the direction of transform motion and thereby creating small regions where the transform motion causes the crust to stretch, become attenuated and subside. The Dead Sea, the Sea of Galilee and the Gulf of Aqaba are all instances of such "pull-apart basins" along the fault. North of the northernmost basin the fault zone bends and steps to the east, producing the opposite result: the compression and thickening of the crust, which has raised the Palmyran Folds. In this way some of the northward motion of the Arabian plate with respect to Europe has been absorbed by convergence and shortening within the continental crust.

It is convergent plate-boundary systems that generate most of the continental crust (see Figure 1.6). Among the three types of boundaries they are the most complex type, and in addition they deform the continental crust across the widest region. In the most usual configuration of a convergent boundary system a plate of oceanic lithosphere is subducted under an overriding plate of either oceanic or continental lithosphere. At increasing distance from the zone of subduction the overriding plate commonly shows a sequence of geologic features: first an accretionary wedge of folded and faulted sedimentary rocks and fragments of oceanic crust scraped from the top of the downgoing plate; then a topographic maximum (an "outer-arc high") formed by the most

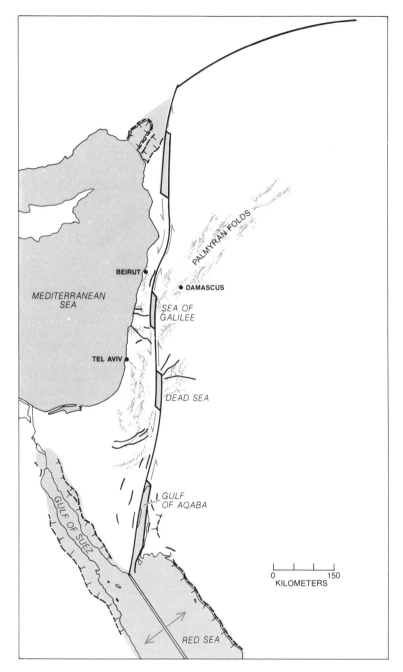

Figure 1.5 TRANSFORM PLATE BOUNDARY, where two plates slide past each other, is exemplified by the Dead Sea fault zone in the Middle East. The crust at the east of the fault is moving north with respect to the crust at the west, and the relative displacement, which amounts to about 105 kilometers in the southern part of the zone, has opened a number of gulfs and seas, of which the Dead Sea is one. In addition parts of crust have shortened to form the Palmyran Folds.

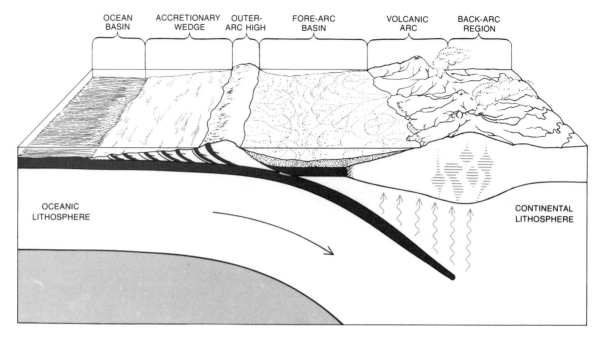

| OCEAN BASIN | ACCRETIONARY WEDGE | OUTER-ARC HIGH | FORE-ARC BASIN | VOLCANIC ARC | BACK-ARC REGION |

OCEANIC LITHOSPHERE

CONTINENTAL LITHOSPHERE

Figure 1.6 CONVERGENT PLATE BOUNDARY, where two plates collide, is marked by a characteristic sequence of geologic features in the overriding plate. In the most usual configuration oceanic lithosphere (crust and upper mantle) is subducted under continental lithosphere. Fragments of oceanic crust and sedimentary rock scraped from the subducted plate form an accretionary wedge and an outer-arc high. Next comes a fore-arc basin, which accu- mulates sediments from the adjacent elevations, and then a volcanic arc, the most characteristic feature resulting from subduction. Some of the magma rising from the subduc- tion zone solidifies in the crust. The back-arc region be- hind the volcanoes may show convergence (such as crustal faulting and folding) or divergence (such as crustal thin- ning and subsidence).

elevated parts of the accretionary wedge; then a fore-arc basin, which accumulates sediment from the adjacent elevations, and finally a volcanic arc, the most characteristic feature, fueled by magma rising from the subducted plate and the mantle just above it. If the overriding plate is oceanic litho- sphere, the geologic features form what is called an island arc. If it is continental lithosphere, they form a continental-margin volcanic arc.

Behind the volcanic arc the overriding plate may be extended or compressed. Alternatively it may be relatively passive. If the plate is in exten- sion, structures similar to the ones at divergent plate boundaries can form. A broad zone of stretching may thin the lithosphere and the crust, forming depressions such as the Aegean Sea of the eastern Mediterranean. If the extension of continental crust behind the arc proceeds until new oceanic crust is formed, a marginal sea will result. It will intervene

between the greater part of the continental mass and the newly rifted fragment of continental crust, just as the marginal sea called the Sea of Japan intervenes between Asia and the islands of Japan. If the overriding plate is in compression, folds and faults arise in belts to accommodate the shortening and thickening of the crust behind the arc (see Fig- ure 1.7). In the Andes such features are found more than 800 kilometers behind the subduction zone.

At convergent zones new material from the man- tle is added to the crust. In particular the subduction of oceanic lithosphere carries some ocean-floor sed- iments and the uppermost part of the oceanic crust downward into the mantle. The sediments and the crust contain water, and the water reduces the melt- ing temperature of certain components of the sub- ducted material. It also reduces the melting temper- ature of certain components of the mantle of the overlying plate. In short, the subduction of oceanic lithosphere causes partial melting at depth. The

Figure 1.7 BRITTLE DEFORMATION is characteristic of the rocks at shallow levels in a belt deformed by plate convergence. Hence sheets of rock tend to be thrust one over the other for tens or hundreds of kilometers. This photograph was made in the Spring Mountains of south-ern Nevada. The dark rocks are Cambrian limestones that are 550 to 500 million years old. They have been thrust from right to left over Jurassic sandstones lighter in color, which are 200 to 175 million years old. A well-defined thrust fault some 30 kilometers long marks their interface.

melted igneous material rises into the overlying rock. There it may cool and crystallize to form plutons: large subterranean igneous bodies. It may also reach the surface as lava or as explosive volcanic products such as pumice and ash. It is enriched in the elements common in continental crust; thus the partial melting advances the chemical differentiation of the outer part of the earth.

Geochemical studies of the igneous products show that they have had a complex, multistage history before coming to rest. In many instances the igneous rocks have been contaminated by contact with older crustal rock, so that not all their volume represents new material derived from the mantle. Indeed, some igneous rocks are derived entirely from the melting of continental crust; hence they add nothing to the volume of the crust. The question of how much of the material added to the continents through igneous intrusion is new and how much is recycled is still unresolved.

In any case the igneous intrusions increase the temperature within the lower crust, and the increase enhances the ability of the crustal rock to lose brittleness and deform in a ductile way. The structures in this ductile part of the crust may thus form large, complex folds. Moreover, the preexisting rocks may recrystallize into rocks with new mineral assemblages. The deformation and recrystallization may obscure and even obliterate the preexisting rock types and deformational patterns, so that it is difficult to elucidate the origin and evolution of the older rocks when erosion uncovers them at the surface of the earth. In general the regions of ductile deformation in the continental crust grade upward and laterally into regions of brittler deformation where the temperature remained lower.

Inevitably the convergence of plates leads to collisions between island arcs and continents. The arcs (along with oceanic islands and plateaus) are a transitional type of crust thicker and less dense than oceanic crust but not as thick and not as "light" as most continental crust. Nevertheless, the arcs, like continental crust, tend to resist subduction. Thus the elimination (by subduction) of the oceanic crust between an arc and a continent that are on converging plates leads to their becoming sutured together. The convergent motion during the collision may be perpendicular to the convergent boundary or it may be oblique to the boundary and have a component of transform motion. Where the motion is oblique the deformation within the collision system will have the characteristics of both convergent and transform boundaries.

The tectonic events at Papua New Guinea exemplify the geologic evolution caused by the collision of island arcs and continents (see Figure 1.8). Here the convergence of the Australian plate and the Asian plate over the past 40 million years has driven one island arc or possibly two up over the edge of the Australian plate. The rock assemblages of the arc (or arcs) have been slivered and foreshortened, along with the assemblages of the ancient margin of Australia. Indeed, the northern edge of Australia has been shortened and thickened as far as 300 kilometers from the site of the collision. The disruption of the arc and of the continent has therefore been severe, but not so severe that the original relations among the rock assemblages cannot be deciphered.

Another type of geologic evolution is found where two continents collide along a convergent plate boundary. Such collisions are occurring today along the Alpine-Himalayan chain, where the Indian plate, the Arabian plate and the African plate are each colliding with the Eurasian plate (see Figure 1.9). In the eastern Mediterranean the collision zone is more than 500 kilometers wide. There the geologic evidence suggests that several small fragments of continental lithosphere were swept together between the converging plates. The "buoyant" fragments stayed at the surface while oceanic tracts were subducted. The continued convergence in the region has deformed both the fragments and the margins of the plates, so that the collision system now extends across a broad zone.

One of the characteristics of collision systems, particularly those between continents, is subhorizontal decoupling, a process in which crustal sheets 10 to 20 kilometers thick slide over one another for tens or hundreds of kilometers. Such displacements stack and thicken different parts of the crust into a

series of irregularly deformed and folded sheets, so that the rock assemblages and structures at depth in the crust cannot be predicted from the assemblages and structures exposed at the surface. Another characteristic is that the convergence, which typically occurs along irregular boundaries in crust that is very anisotropic, causes complex motions of small crustal fragments within the convergent system. These local motions may be divergent, transform or convergent.

The most spectacular example of a convergent plate-boundary system active today is in Asia, where Peter Molnar of the Massachusetts Institute of Technology and Paul Tapponnier of the French National Scientific Research Council (CNRS) were the first to recognize that deformation extends across a region 3,000 kilometers wide. The collision some 50 million years ago between the Indian plate and the Asian plate represented a collision between the continental lithosphere of India and that of Asia. Since then a continued convergence that may total more than 2,000 kilometers has been absorbed principally by strain in the Asian plate. Broadly speaking, the Asian plate has absorbed the massive intracontinental deformation by compressional, transform and extensional faulting along young belts of deformation that generally follow older belts resulting from the activity of more ancient plate-boundary systems. To put it more simply, Asia has shortened longitudinally and extended latitudinally to accommodate the northward movement of India.

Meanwhile the northern edge of the Indian plate has broken into several gently dipping slabs whose pileup has thickened the crust and formed the Himalayas. The faulting in Asia extends nearly 3,000 kilometers from the collision boundary. Igneous activity in parts of the collision zone suggests that the deeper parts of the crust remain very hot today, creating an environment where rocks are recrystallizing and undergoing ductile deformation. Hence the rock assemblages and structures formed at earlier times are now being "overprinted."

From studies of the youngest deformed belts, such as the Alpine-Himalayan belt, it is plain that plate boundaries evolve rapidly. Island arcs can be created, travel thousands of kilometers and collide with continents in only a few tens of millions of years. Small continental fragments can be rifted and collide with continents over similarly short spans of time. Thus the deformed belts that make up the

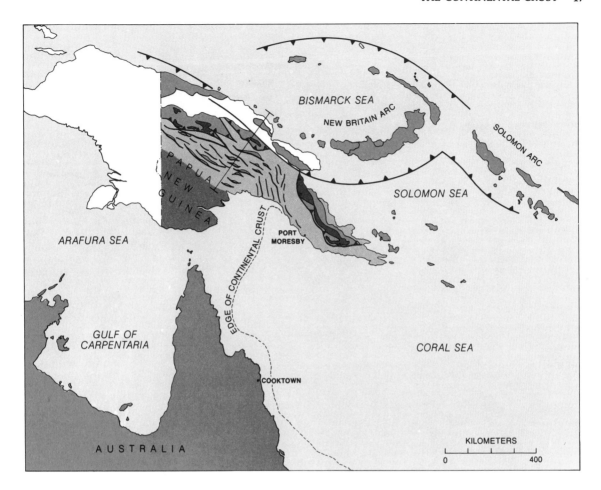

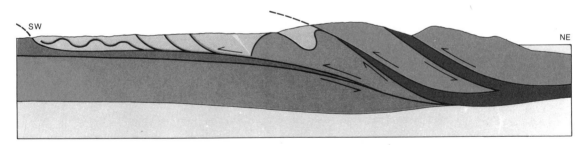

Figure 1.8 COLLISION OF ISLAND ARCS with the northern part of the Australian continental crust over the past 40 million years has sutured some arcs (*green*) and also some remnants of oceanic crust (*gray*) onto the northern part of Papua New Guinea. The arcs were probably formed by the subduction of the Australian plate under the Pacific plate; then the convergence of the plates carried Australia into the zone of the subduction. The continued convergence has now folded and faulted sedimentary rocks along the margin of Australia (*purple*) and also the ancient crust of Australia itself (*orange*). Darker shades signify rock that escaped deformation. The cross section is schematic.

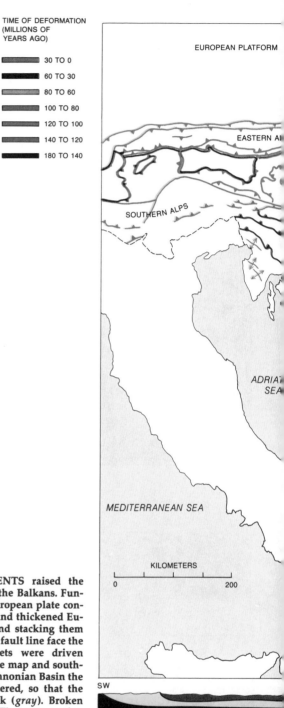

TIME OF DEFORMATION
(MILLIONS OF
YEARS AGO)

30 TO 0

60 TO 30

80 TO 60

100 TO 80

120 TO 100

140 TO 120

180 TO 140

EUROPEAN PLATFORM

EASTERN A

SOUTHERN ALPS

ADRIAT
SEA

MEDITERRANEAN SEA

KILOMETERS

0   200

SW

Figure 1.9 COLLISION OF CONTINENTS raised the mountain chains of eastern Europe and the Balkans. Fundamentally the African plate and the European plate converged, and the convergence shortened and thickened Europe by faulting its rocks into sheets and stacking them one above the other. Small barbs on each fault line face the overriding sheet. The overriding sheets were driven northward in the northeastern half of the map and southward in the southwestern half. In the Pannonian Basin the crust has extended, thinned and foundered, so that the stacking is covered by sedimentary rock (gray). Broken lines indicate outcrops of deeper rock. The cross section suggests the complexity of the stacking; the colors employed differentiate sheets of rock, which are highly deformed.

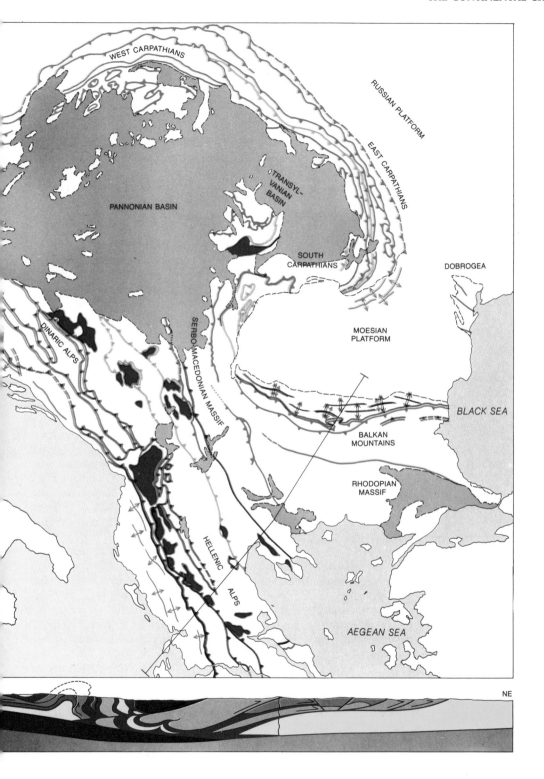

WEST CARPATHIANS

RUSSIAN PLATFORM

EAST CARPATHIANS

TRANSYL-
VANIAN
BASIN

PANNONIAN BASIN

SOUTH
CARPATHIANS

DOBROGEA

MOESIAN
PLATFORM

DINARIC ALPS

SERBO-MACEDONIAN MASSIF

BLACK SEA

BALKAN
MOUNTAINS

RHODOPIAN
MASSIF

HELLENIC
ALPS

AEGEAN SEA

NE

underpinning of the continents represent a long, complex history of superposed plate-boundary systems.

The end of such a history probably comes when a region gets to be so remote from plate-boundary activity that it is no longer under its influence. In many instances this happens when an ocean basin is finally closed by a continent-continent collision. Parts of the Ural Mountains of the central U.S.S.R. are an example. First, about a billion years ago, two continental masses were rifted apart and a large ocean basin was created between them. The subsequent closing of that ocean generated island arcs, which collided with the continents at various times until the sequence was ended by the collision of two continents 250 million years ago. Much of the deformed belt resulting from that collision (namely the chain of the Urals) lies far enough from any younger plate-boundary activity for it to have suffered no further deformation.

This is not to say the entire belt is immune. Rifting in the south has created oceans younger than 250 million years, some of which have closed and created new mountain belts; hence the southern extension of the Urals has been overprinted by the Alpine-Himalayan belt. Moreover, the northern extension of the Urals reaches the Arctic Ocean, where future plate-boundary activity may be in store. Similar histories can be read in the geology of essentially all the continents in the crosscutting of older deformed belts by younger ones.

Of course, some histories are easier to read than others. In the youngest deformed belts the timing of tectonic events can be distinguished with an accuracy of better than one million years. In the older belts the timing becomes poorer with increasing age. The relative timing of many events can be determined, but the contemporaneity of events over large areas is difficult to establish. Thus it becomes a challenge to make accurate reconstructions of ancient plate-boundary systems. In addition the older belts are likelier to be ones where the once continuous locus of deformation has been disrupted. For example, the continuations of the ancient deformed belts in Australia are now found in India, Africa, Antarctica and South America.

The erosion of old deformed belts in regions where convergence has thickened the crust offers a valuable opportunity to study rocks that were once at deep crustal levels. It is remarkable just how deep erosion can reach. Erosion cuts into the terrain (a process taking tens of millions of years, even several

tens of millions), progressively uncovering deeper levels, and since the crust is "buoyant," the removal of material from the top of it causes the remaining crust to rise. In effect it is timber pushed toward a saw. In this way rocks from depths as great as 30 or 40 kilometers come to be exposed at the surface.

By the time the rocks are exposed the plate-boundary forces that shaped them have been long inactive. Still, the examination of the rocks enables geologists to infer the processes, the temperatures and the pressures that existed as the rocks deformed, and from such results a picture of the three-dimensional response of the continental crust to plate-boundary activity can be constructed. Only the deepest crustal levels excape exposure by erosion. Studies of inclusions in igneous rocks, chemical studies of igneous rocks and geophysical studies suggest that much of the deepest crust has a composition not unlike the composition of the shallower crust, except that the deep rocks are recrystallized so that their mineral assemblages are ones that are stable at high temperatures and pressures.

Some of the older deformed belts in the continents, including most of the Archean belts (the belts from 2.5 to 3.8 billion years old), have been difficult to interpret as plate-boundary systems analogous to modern ones. To be sure, the types of rocks in the older belts are similar to the types found in modern convergent systems. Their arrangements and structures, however, are somewhat different. Typically the older belts consist of bodies of volcanic and sedimentary rock, irregular or elongated in shape, along with large expanses of intrusive granitic and deformed metamorphic rocks that include mineral assemblages formed at high temperatures and pressures. The volcanic and sedimentary bodies resemble those found in modern island-arc settings. The only notable differences are that basaltic rocks are more abundant in the older belts and that some of the basalts are richer in magnesium and poorer in silicon than most modern basalts. On the other hand, many features of the younger belts appear to be missing in older ones. Thick, widespread sequences of shallow-water sedimentary rocks of the type that develop on modern passive continental margins are one example. Widespread vertical stackings of sheets of crustal rock that have been thrust one over the other are a second example.

Investigators of the Archean belts have taken two points of view: that the Archean belts are the result of plate motions whose geometry and intensity differed from those of modern plates, or that plate tectonics did not operate in the Archean, so that a mechanism of geologic evolution not observed today must be invoked instead. Although each view has its adherents, a modification of plate-tectonic theory can also be proposed. Perhaps the volcanic and sedimentary bodies in the Archean belts represent island arcs and their associated fore-arc basins and marginal seas, all of which were swept against small colliding continental nuclei. Larger, continental masses developed progressively, so that by about 2.5 billion years ago some orogenic belts began to take on a more modern appearance.

The rate at which continental crust has been formed remains a major question. On the one hand, the examination of orogenic belts indicates that many of them include a large amount of material derived from older belts by reworking or remelting. In addition many belts include rocks derived directly from the mantle by volcanic activity. Further still, many belts incorporate oceanic lithosphere. On the other hand, evidence suggests that continental material may sometimes be lost by being taken into the mantle. The studies show that orogenic belts vary widely in the balance of these processes. For example, the orogenic belts that were formed in north-central Canada between 2.5 and 1.8 billion years ago are proving to include much crust older than 2.5 billion years. In contrast, an orogenic belt of the same age in the southwestern U.S. contains little if any older crust. The studies do seem to indicate that the volume of continental crust has increased with time.

One of the very oldest Archean belts, about 3.8 billion years old, is in the continental crust of Greenland. Its rocks are sedimentary and igneous, and material in the sedimentary rocks has been derived in part from some older continental rocks. No direct evidence from that earlier era has been discovered, however, and so there remains a gap of 800 million years beginning 4.6 billion years ago, when the solar system is thought to have formed. In the later part of this gap the moon was intensely cratered; surely the earth was subjected to a similar bombardment. No signs of it remain. They were probably erased by the dynamic processes that have continuously created and reworked the continental crust.

Clearly the earth is an evolving body whose distribution of heat controls the motions, thickness and ductility of the lithosphere and the generation of igneous and metamorphic rocks. The generation of heat by radioactivity in the earth was probably about three times greater in the Archean than it is today. Temperature gradients in the earth were probably greater too, and that may help to explain at least some of the differences between Archean belts and younger ones. Until the variables that affect the formation of orogenic belts are better understood, the Archean belts will remain a major challenge to the understanding of how the continents came into existence.

The Oceanic Crust

It is created and destroyed in a flow outward from midocean ridges to subduction zones, where it plunges back into the mantle. Currently it is being opened to view by submersibles and novel instrumentation.

. . .

Jean Francheteau
September, 1983

From the point of view of the earth scientist our planet probably should be called Ocean rather than Earth, not only because 70 percent of it is covered by water but also because 60 percent of its solid surface is covered by the thin crust that is manufactured in a unique geologic mill at midocean. The first piece of the oceanic crust to be identified was recovered by the British cable-laying steamship *Faraday* in 1874. The *Faraday* was sailing in the North Atlantic on a mission to mend a broken transatlantic telegraph cable. The cable had broken in 2,242 fathoms of water where it passed over a large rise in the ocean floor later named the Faraday Plateau. The *Faraday* was equipped with a large grapnel for lifting cables from the ocean floor.

According to Marshall Hall, an English geologist, "whilst engaged in grappling for the broken telegraph cable the ship caught the strong claws of the grapnel in a rock, which resisted with a strain of about 27.5 tons, to which any but a rope of marvelously perfect manufacture would have yielded. As it was, the rock gave way and a lump of black basalt came up weighing 21 lb. This mass shewed signs of having been torn off." The basaltic rock was brought back to England, and in 1876 Hall and J. Clifton Ward examined and described it.

In the century following the finding of this first piece of it the oceanic crust has come to have a central position in earth science. It is now known that the Faraday Plateau is one segment of a 59,000-kilometer system of ridges that girdle the earth under the sea. The midocean ridge has great significance in the theory of plate tectonics, which transformed the earth sciences in the late 1960's and early 1970's. The ridge marks the boundary between two rigid plates that are supported by the underlying mantle of the earth. At the ridge the plates separate slowly and the underlying rock rises to fill the gap, melting as it does so. Thus a few square kilometers of new oceanic crust is formed each year at the crest of the midocean ridges. The crust that is formed in this way is profoundly different from the crust of the continents. It is an order of magnitude younger on the geologic time scale than the continental crust and has a quite different composition from the continental masses.

As a result of many decades of observation that were unified in the plate-tectonic hypothesis it is now known where and roughly how oceanic crust is formed. The detailed structure of the crust, however, is much less well known. Therefore in the late 1970's and early 1980's the work on the oceanic crust has turned from global theory-making to investigating details of structure and composition. The oceanic crust now appears to be much more diversified both in its topography and in its layered structure than had previously been thought. The refinements in understanding the oceanic crust that have been achieved are due in large part to novel techniques of observation of the sea floor, which remains one of the least accessible parts of the planet's surface.

Satellite measurements of the gravity field over the oceans are yielding an improved picture of the general topography of the ocean bottom. The detailed topography is being mapped by new sonic methods. The layers of the crust are being probed by deep drilling, by new seismic methods and by measurements of electrical conductivity. The large body of new information that is being made available by such techniques is rapidly changing the accepted notion of the oceanic crust. The work is by no means complete, but within a few years we should have a substantially better picture of the thin crust that covers the larger part of the earth's solid surface.

In plate-tectonic theory the crust and the upper mantle of the earth are divided into the lithosphere, or strong layer, and the asthenosphere, or weak layer. The lithosphere includes the crust and part of the upper mantle. In the ocean the crust is on the average from five to seven kilometers thick; away from the ridge crest the lithosphere is about 100 kilometers thick. The lithosphere is broken up into a set of fairly rigid plates that are much like rafts floating on the less rigid material of the asthenosphere.

The plates move at a rate of a few centimeters per year with respect to each other, and the boundary between two plates can be described according to the relative plate motion. At divergent boundaries the plates separate. At convergent boundaries the plates move toward each other, and one plate generally plunges under the other and into the asthenosphere in the process called subduction. At transform boundaries the plates slide past each other. The spreading center of the midocean ridge, where molten rock from the mantle is injected into the crust, is a divergent plate boundary.

The mantle below the spreading center is composed mainly of peridotite, a type of rock consisting mainly of the mineral olivine, which in turn consists mainly of magnesium, iron, oxygen and silicon but is poor in silicon compared with the rocks of the crust. The separation of the plates at the ridge reduces the downward pressure on the mantle rock below. Hence a part of the mantle begins to move upward; the zone of upwelling extends from a depth of 50 to 70 kilometers to the base of the crust. The decompression of the mantle material is adiabatic, meaning that it takes place without loss of heat, and under such conditions the peridotite begins to melt as it rises.

Not all the peridotite melts on the way to the surface. In general the basaltic liquid injected at the spreading center is formed by the melting of 10 to 20 percent of the upwelling mantle rock. The melt collects in a magma chamber at the base of the crust, where it separates through crystal fractionation into fractions of different composition (see Figure 2.1). The fractions lie above a residual solid that has the composition of peridotite. Within the chamber slow cooling and crystal fractionation result in the formation of gabbro, a type of rock incorporating in addition to olivine the mineral plagioclase, which consists mainly of silicon, oxygen, sodium and calcium. Gabbros and other rocks that are formed by accumulation make up the lower layer of the crust. The basaltic liquid in the upper part of the chamber reaches the surface through a system of vertical passages. Once at the surface, the liquid flows down the slopes of the ridge and hardens into sheets of the rounded forms known as pillow lavas. Which of the forms results depends on the slope of the ridge and the rate at which the lava is extruded. Moreover, as the plates move apart the ascending magma hardens into a series of dikes: massive vertical sheets.

The injection of magma, or molten silicate liquid, plugs the gap left by the moving apart of the plates. The plates continue to diverge, however, and the plug is ultimately rifted open. A new cycle of asthenospheric upwelling, peridotite melting, separation in the magma chamber and extrusion begins. Meanwhile the crust formed in the previous round of upwelling is moving outward from the spreading center. As the crust moves outward it is modified. Tension exerted by the continuing plate motion can result in a series of fissures and faults running parallel to the strike, or longitudinal axis, of the ridge crest (see Figure 2.2).

As the crust cools many small cracks appear in its upper layers. Seawater penetrates the cracks and the fissures caused by the tension in the crust. The water flows downward into the crust, is heated and

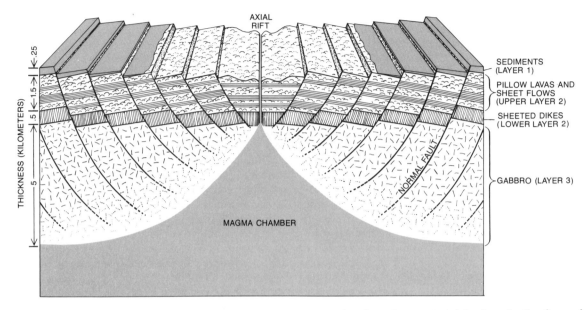

THICKNESS (KILOMETERS)

.25
1.5
.5
5

AXIAL
RIFT

SEDIMENTS
(LAYER 1)

PILLOW LAVAS AND
SHEET FLOWS
(UPPER LAYER 2)

SHEETED DIKES
(LOWER LAYER 2)

NORMAL FAULT

GABBRO (LAYER 3)

MAGMA CHAMBER

Figure 2.1 SPREADING CENTER is where magma is injected into the crust, as shown in this cross section of the midocean ridge. The magma forms as the lithospheric plates diverge and mantle rock rises and melts with the decrease of pressure. It collects in a chamber below the spreading center. Within the magma chamber the rock gabbro crystallizes. At the top of the chamber the magma rises as the plates diverge and cools as vertical dikes. At the surface lava flows out and hardens in the form of sheets and "pillows." As the new crust moves away from the spreading center, a layer of sediments is deposited on it. The crust also cracks along normal faults, which run parallel to the ridge crest. Hence the mature crust has a layered structure from the top down: sediments, sheet and pillow lavas, dikes and gabbros.

rises to the surface of the crust again. Such hydrothermal circulation efficiently leaches water-soluble compounds out of the rock. Metallic elements, which readily form ionic complexes, are leached with particular efficiency. The leached elements are carried upward and the hot seawater with its burden of metals is expelled into the ocean by vents near the ridge crest. The discovery of the vents and the exotic biological communities that cluster around them was one of the most exciting findings of oceanic science in the late 1970s (see Figure 2.3).

The hot circulating water and shallow heat sources lead to metamorphic changes in the lower crust and rapid chemical changes in the upper crust. In addition, as the crust moves outward a layer of sediment is deposited on its upper surface. The sediments consist largely of the remains of tiny oceanic plants and animals. The type of sediment that is deposited and the depth of the sediment cover therefore depend largely on the biological activity of the ocean.

Thus at some distance from the ridge crest the oceanic crust has the following vertical structure from the ocean floor downward. At the top is a layer of sediment perhaps .5 kilometer thick. Under the sediment is a layer known as the oceanic basement made up of interspersed sheet flows and pillow lavas with the underlying complex of vertical dikes. The basement can be two kilometers thick and is heavily fractured and altered by hydrothermal circulation. The third layer, the oceanic layer, is made up of the gabbros that solidify out of the basaltic melt in the magma chamber. The gabbros can undergo considerable metamorphism as they move away from the ridge. The oceanic layer is perhaps five kilometers thick.

This model of the layered structure of the oceanic crust and how it is formed is based on marine geophysical data, on studies of rocks from the ocean floor, on observations of fragments of oceanic crust emplaced in mountain belts on dry land and on conjectures. The rest of this article will be devoted to examining some of the unsolved problems pre-

Figure 2.2 FISSURE CUTS THROUGH THE SURFACE of the oceanic crust near the crest of the East Pacific Rise. Such fissures are caused by plate motion, which generates lateral stresses, and also by the contraction of the crust. The fissures nearest the ridge crest are active sources of lava flow and are often concealed by the flows. Those farther from the ridge crest, like the one shown, no longer expel lava. (Photo taken in 1978 during the first exploration of the East Pacific Rise by the manned submersible *Cyana* in an expedition led by J. Francheteau.)

sented by that model. It will become clear that the simple picture of the crust that was thought to be accurate until quite recently is rapidly being made more complex by new findings.

Since the first deep-sea soundings were made in the second half of the 19th century it has been known that the ocean floor lies much deeper below the sea surface than the continents rise above it. Away from the continental margins, which are not made up of oceanic crust, the ocean is on the average 3.7 kilometers deep. The great depth of the ocean and the sediment cover on the oceanic crust make the crust difficult to observe with most geologic techniques. In the past decade, however, several technical advances have greatly increased the accumulated knowledge of the oceanic crust.

At the most general level the methods of observation can be put in two groups: those that reveal the topography of the surface of the crust and those that penetrate below the ocean floor to gain information about the composition and vertical structure of the crust. In the studies of topography much

interest is focused on the midocean ridge that marks the spreading center. The ridge is a long linear upswelling with a gradual slope. The average depth at the crest of the ridge is about 2.5 kilometers, and the ocean bottom slopes away on both sides to a depth of from five to six kilometers.

As the oceanic crust is carried away from the ridge crest it cools and contracts. In this process the lithospheric plate can be thought of as "floating" on the asthenosphere. Consider a block of wood floating in a barrel of water. If the block is not perturbed, it will come to rest in the water at a level corresponding to a balance between the downward force of gravity and the buoyancy of the wood. The buoyancy and hence the point of equilibrium depend on the density of the wood. Similarly, in the absence of perturbing factors the lithospheric plate floats on the asthenosphere at a depth corresponding to what is called isostatic equilibrium: the level at which the weight of the lithosphere is balanced by the pressure from the mantle. The level to which the plate sinks depends on the density of the rock in the

Figure 2.3 BLACK SMOKERS are vents through which hot water is expelled from the crust. The particles that give the plume its dark color are sulfides that have been dissolved out of the crustal rock. Near the midocean ridge seawater penetrates cracks in the newly formed crust. The water is heated and propelled upward through the vents. (Photo taken on the East Pacific Rise from the manned submersible *Cyana* during an expedition led by R. Hékinian.)

lithospheric column, a thin vertical section through the lithosphere.

As the oceanic crust cools and contracts its density increases and hence it sinks deeper into the asthenosphere. The depth to which the crust sinks has been shown to vary with the square root of its age. Crust that is two million years old lies at about three kilometers, crust that is 20 million years old lies at four kilometers and crust that is 50 million years old lies at five kilometers. Charting the topography of the ocean bottom can therefore provide an estimate of the age of the crust (see Figure 2.4).

Topographic maps can yield other significant information on the movement of the plates. Because lithospheric plates are rigid bodies, when two plates separate, their motion can be described in relation to a point on the earth's surface that is referred to as the pole of rotation. (The pole of rotation should not be confused with the earth's geographic or magnetic poles; it is significant only in relation to plate mo-

tion. Furthermore, the pole describing the relative motion of a pair of plates can shift several times over the history of the plate's interaction.) Continuous segments of spreading axis along the ridge crest define great circles that pass through the pole of rotation much as a meridian, or line of longitude, passes through the geographic pole.

As two plates move around their pole of rotation transform faults caused by stresses in the plates

Figure 2.4 EAST PACIFIC RISE appears as a raised area running roughly north and south in this topographic map showing the Pacific Ocean off the coast of South America. Brown designates shallow regions, yellow intermediate regions and green deep regions. Each ridge marks the place where two lithospheric plates diverge; the East Pacific Rise is the boundary between the Pacific plate and the Nazca plate. The axis of the rise is crossed by many of the large faults called transform faults. (Map based on depth-soundings data compiled by the U.S. Naval Oceanographic Office and plotted and colored by computer at the Lamont-Doherty Geological Observatory.)

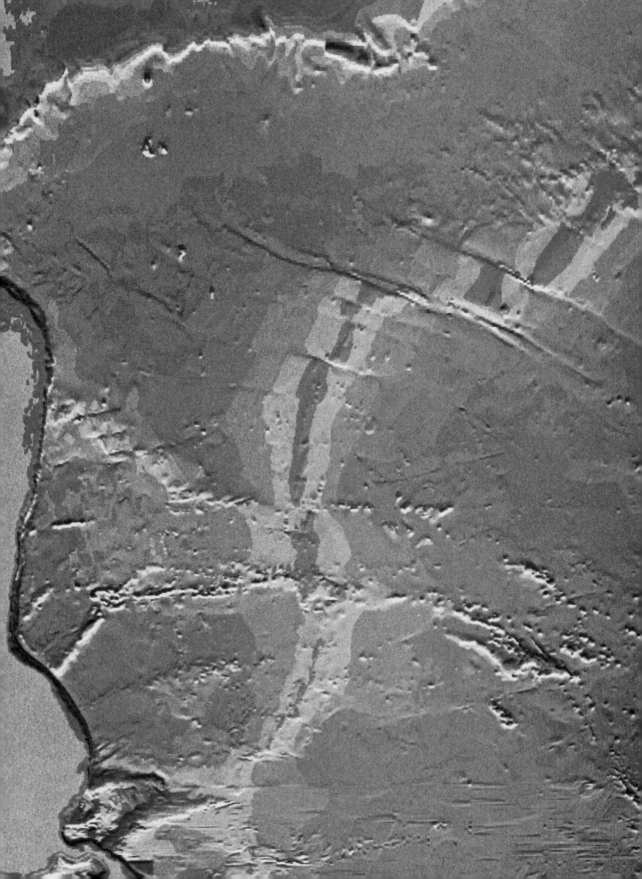

form across the strike of the ridge. Other transform faults are caused by the irregularity of the initial rupture within a continent that created the ocean basin. When the land masses separate, the breaks parallel to the direction of plate motion develop into transform faults; the breaks perpendicular to the plate motion develop into spreading centers. At a transform fault the axis of the ridge is offset, and the parts of the plates slide by each other in opposite directions (see Figure 2.5).

The offsets at the transform faults are roughly parallel to the direction of relative plate motion. The offsets have the effect of keeping the axis perpendicular to the direction of spreading. On the Mid-Atlantic Ridge transform faults can be as close together as 50 kilometers. Away from the ridge, outside the offset, the parts of the plate are not in motion with respect to one another. In this region structures called fracture zones mark the position of the transform faults; the fracture zones extend from the ridge crest like ribs from a spine. Between the transform faults are many smaller faults that are also caused by plate motion.

The precise position of the faults and fracture zones on the ridge holds an excellent record of the kinematics of the plates: the history of the relative plate motions. Mapping techniques that enable geologists to identify the position of these transverse features can therefore serve as the basis for reconstructing the history of the plates, by a method analogous to projecting a reel of motion-picture film in reverse.

One of the most promising tools for reconstructing the kinematics of the plates is a kind of mapping that has been named geotectonic imagery by its developer, William F. Haxby of the Lamont-Doherty Geological Observatory of Columbia University. Geotectonic images come from data gathered by *Seasat*, a satellite launched by the National Aeronautics and Space Administration in June, 1978. *Seasat* was equipped with a radar altimeter capable of measuring the height of the sea surface to within five to 10 centimeters. The instruments on board the satellite failed prematurely after three months of operation, but by then the device, orbiting at an altitude of 800 kilometers, had surveyed the world's

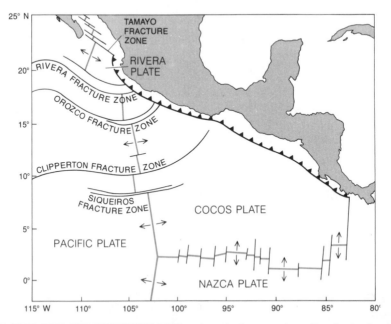

Figure 2.5 EAST PACIFIC RISE OFF SOUTHERN MEXICO has a complex geometry that includes many faults across the strike, or main axis, of the ridge. The axis of the rise runs along the boundary between the Pacific plate and the Cocos plate; arrows indicate the direction of plate motion. Fracture zones mark the points where major transform faults cross the strike. At the fracture zone the ridge axis is offset. Within the offset the direction of movement of the crust on one side of the fault is opposite to that of the crust on the other side. Between the fracture zones are many smaller faults.

oceans between 72 degrees north and 72 degrees south.

The data from the three months of *Seasat* operation are currently being processed to yield dramatic images of the ocean floor (see Figure 2.6). The principle employed to convert the information on sea height into topographic maps is an intriguing one. The main cause of the spatial variation in the height of the sea surface is variation in the gravity field at sea level: the ocean tends to "pile up" where the gravity field is high and do the opposite where the gravity field is low.

The differences in gravitation are measured in relation to the ellipsoid. The ellipsoid is a mathematical figure corresponding to what the average sea surface would be if the mass of the earth were distributed in a radially symmetrical way.

The mass of the upper layers of the earth under the oceans is not, however, distributed in a radially symmetrical way. Hence the sea surface does not follow the ellipsoid. Instead it follows an irregular figure called the geoid. The difference between the geoid and the ellipsoid at any point on the ocean corresponds to the gravitational anomaly there. Where the geoid is higher than the ellipsoid there is a positive gravitational anomaly. Where the geoid is lower there is a negative anomaly.

What could account for the differences in the gravity field? Since the gravitational force exerted by a body is proportional to its mass, the presence of large topographic features such as seamounts, or undersea mountains, rising from the ocean floor is associated with positive gravitational anomalies; depressions or valleys in the floor are associated with negative anomalies.

In the *Seasat* maps of the ocean floor the fracture zones, volcanic seamounts and deep trenches near subduction zones stand out as if the ocean had been drained away. Their spectacular vistas aside, geotectonic images are being put to significant geophysical uses. For example, by providing detailed maps of fracture zones the geotectonic images are helping in the reconstruction of the direction of plate motion in the past.

The geotectonic images do not correspond exactly to the topography of the sea floor. Although seamounts, fracture zones and trenches are rendered with great clarity, one of the main spreading centers in the Pacific, the East Pacific Rise, appears as an insignificant swell. The reason for the discrepancy can be understood by considering again the block of wood. Imagine the block is a topographic feature on the crust and *Seasat* is flying over it comparing its gravity field to the fields around it.

When the block floats, part of it extends above the surface; hence there is an additional mass at that point on the surface and a positive gravitational anomaly would be expected there. The block is less dense than the fluid in which it floats, however, and so below the surface there is somewhat less mass than there is at the nearby points where no blocks are floating. When the block is at flotational equilibrium, the mass added above the surface and the mass subtracted below the surface are equal. Therefore under conditions of isostatic equilibrium in the crust a much smaller gravitational anomaly is observed than would be expected on the basis of the topography alone. The topographic feature is thus almost invisible to *Seasat*.

The midocean ridges are approximately in isostatic equilibrium, and so they appear as relatively small upswellings, much smaller than they are in reality. Other features are out of equilibrium and appear quite clearly. Consider a seamount on a piece of old oceanic crust. The crust has cooled and contracted and is quite rigid; hence the seamount cannot sink down far enough to reach equilibrium at that point. Instead the crust is depressed only slightly under the seamount but the shallow depression extends a long distance around the mount. Thus the large area that includes the seamount is in equilibrium but the point where the mount sits is not. The subtracted mass is spread over a considerable area of the crust, whereas the added mass is concentrated at the seamount. The result is a positive gravitational anomaly at the seamount and a smaller negative anomaly around it.

Because the Mid-Atlantic Ridge and the East Pacific Rise are almost invisible to instruments that measure the gravity field, other techniques must be employed to map them. The most significant are methods based on recording acoustic pulses with a very high frequency projected from a device on shipboard to the ocean floor. The depth under the ship can be calculated from the travel time of the pulse from the ship to the bottom and back. In addition to the capacity to map spreading centers such sonic methods can provide much greater detail than a satellite can.

Sonic techniques are not new, but recent innovations have made them much more valuable to earth scientists than they were previously. The most sig-

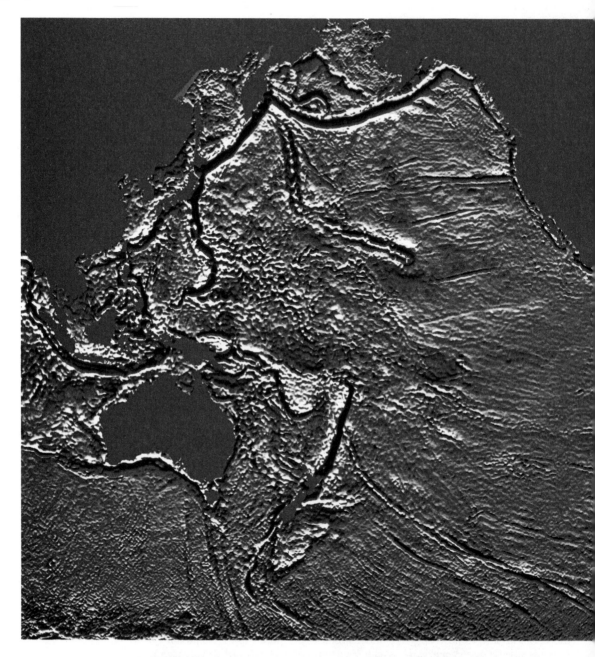

Figure 2.6 GRAVITY-BASED TECTONIC IMAGE of the world's oceans. White areas are shallow (high-gravity) regions, blue areas are deep (low-gravity) regions, and red areas are intermediate. The red-and-white strip between Europe and America is the Mid-Atlantic Ridge. Dark blue lines in the western Pacific are deep sea trenches asso-

nificant innovation is arrays of sonar sources and receivers mounted in a single device. The information from the multiple beams can quickly be combined into a detailed topographic picture. An even more recent innovation is to mount such arrays on vehicles towed at considerable depth behind a ship.

The system called SeaBeam, made by the General Instrument Corporation, has an array of 16 beams,

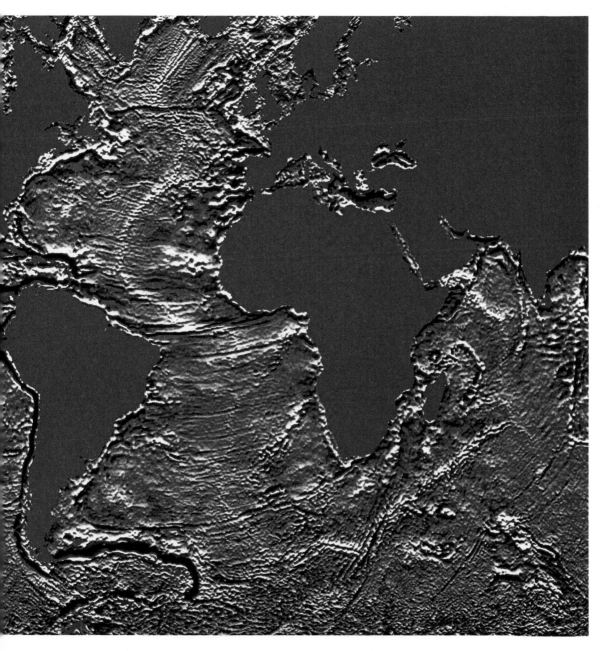

ciated with subduction zones. The bent diagonal line in the Pacific is the Hawaii-Emperor chain of volcanic islands **and seamounts (undersea volcanoes). Hawaii is at its southeast end.**

each with a frequency of 12,000 hertz (cycles per second). Each beam is a little less than three degrees wide. The beams are projected out in the shape of a fan along a line perpendicular to the ship's course, enabling the device to map a swath of sea floor two-thirds the water depth as the ship advances.

High-resolution SeaBeam mapping is a powerful tool for examining the sea floor, particularly the parts of the ridge system where the plates diverge rapidly. The spreading rate at the ridge crest can vary considerably. The Mid-Atlantic Ridge is a medium-to-slow spreading center with a rate of about five centimeters per year. The East Pacific Rise, on the other hand, is a fast spreading center where the rate of divergence can be as high as 15 centimeters per year. At fast spreading centers the features at the crest of the ridge tend to be concentrated in a fairly narrow zone that can be taken in with a single pass of SeaBeam; hence the system is particularly useful for mapping areas such as the crest of the East Pacific Rise.

The axis of the East Pacific Rise runs roughly north-south off the coast of Central America and South America. At its crest is a ridge about 30 kilometers across and 500 meters high. The axial ridge has recently been the subject of intensive work both by earth scientists interested in the fast spreading center and mining companies interested in the sulfide compounds deposited along the ridge by the heated seawater rising from the vents.

The crest of the ridge includes a zone of active volcanism. In the course of dives with the French submersible *Cyana* in early 1982 other investigators and I observed that the area of current volcanism is only one to two kilometers wide, so that it can easily be mapped in one SeaBeam pass. Indeed, about two years earlier a SeaBeam map of the "Great Pacific Highway," the narrow, flat crest of the rise, was made from the French research vessel *Jean Charcot*. Similar maps have since been made by other vessels, and the detailed topography of the East Pacific Rise is becoming clear (see Figure 2.7).

The simultaneous employment of sonar mapping and manned submersibles on the East Pacific Rise has revealed a dramatic topographic pattern that could modify the notion of how the spreading center works. In the northern part of the East Pacific Rise major fracture zones marked by transform faults interrupt the ridge crest every 200 to 300 kilometers. Between the large faults are many smaller faults, some as little as 10 kilometers apart.

Near the transform fault the ridge is quite deep, and it rises to a peak between each pair of faults. The topographic high is usually equidistant from the two faults; the overall swell is about 500 meters over the 200-to-300-kilometer distance. Thus in a profile taken along the strike the ridge has the appearance of a gently rising hill. Since the profile of the ridge transverse to the strike is also that of a gentle rise, the area between a pair of transform faults is shaped like a low dome (see Figure 2.8).

A second set of smaller domes protrudes from the large structure like a row of blisters along the ridge crest. This second set of domes was discovered in multibeam sonar studies by Robert D. Ballard of the Woods Hole Oceanographic Institution, Roger Hékinian of the Centre National pour l'Exploitation des Océans and me on the East Pacific Rise between the Orozco Fracture Zone at 15 degrees north and the Clipperton Fracture Zone at 10 degrees north. The information from the sonar maps was augmented by dives in *Cyana*.

The small domes are bounded by the faults that cut the ridge between transform faults. The small projections are about 100 meters high. It is probable that each dome corresponds to a spreading center: a small area where the manufacture of new crust goes on independent of the adjacent ridge segments. Thus instead of a single great factory the midocean ridge could be a string of small adjacent workshops. Hans Schouten and his colleagues at Woods Hole have suggested that the midocean ridge is indeed such a chain of adjacent spreading cells separated by fracture zones; the cells could be stable for long periods of time. Schouten's hypothesis implies that oceanic crust is not created as a single homogeneous mass but is made in long narrow ribbons laid side by side with fracture zones in between (see Figure 2.9).

Although SeaBeam is the most convenient sonar system for mapping along-strike topography, two other sonar systems can give a more detailed picture of the ocean floor than SeaBeam can. Both are "side-looking" sonars, meaning that the acoustic pulses are projected to the side from devices towed behind the ship near the ocean floor. The Geological Long Range Inclined ASDIC (GLORIA) was developed at the Institute of Oceanographic Sciences at Wormley in England. The sonar transmitters encased in the "fish," a neutrally buoyant housing, emit pulses of sound with a frequency of 6,200 to 6,800 hertz. GLORIA can map a swath of sea floor about 30 kilometers wide.

A side-looking sonar system has one great advantage over a single-channel sonic profiling system operated from shipboard. In the shipboard systems vertical features such as scarps are difficult to detect because they yield little upward reflection of the acoustic beams. In the side-looking system, however, vertical features show up clearly because they offer excellent surfaces for horizontal reflection. Thus at the midocean ridges GLORIA has made plain

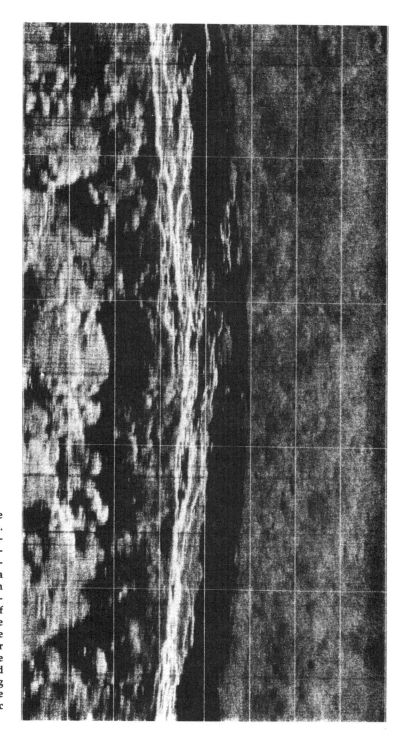

Figure 2.7 AXIAL RIFT delineates the exact center of the midocean ridge. This image of the rift on the East Pacific Rise was made with the sonar system SeaMarc I. A source of high-frequency acoustic pulses is housed in a "fish" that is towed close to the ocean floor behind a ship. The pulse is projected to the side and the intensity of the reflected beam is recorded. The data are later converted into an image of the ocean bottom. In this sonar image Highly reflecting surfaces are light and nonreflecting surfaces and shadows are dark. (Image made during a cruise over the Clipperton Fracture Zone at 21 degrees north in the Pacific led by W. B. F. Ryan.)

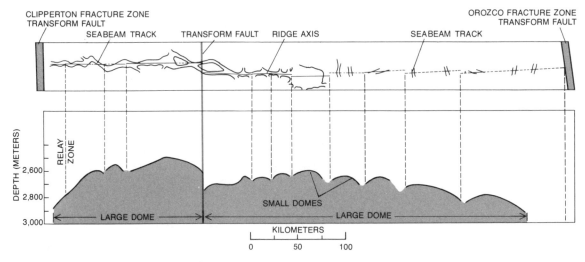

Figure 2.8 SUPERPOSED DOMES at the crest of the East Pacific Rise could give clues to how the spreading center at the midocean ridge operates. The upper panel shows part of the East Pacific Rise in plan view as mapped by the SeaBeam sonar system. The ridge axis is marked by the central line; broken segments stand for regions where the axis had to be guessed at. The axis is marked by active volcanic fissures. Pairs of parallel lines show areas that have been mapped by the SeaBeam sonar system. Ridge offsets mark transform faults and relay zones, where small faults cross the strike. The lower panel shows the same ridge segment in profile. Two large domes run between transform faults, with the peak about halfway between the faults. A series of small domes bounded by relay zones protrude from each large dome.

the pattern of inward-facing scarps along the faults that parallel the ridge axis.

Each side-looking sonar system has its own frequency and arrangement of beams, hence it is best suited for a particular purpose. SeaMarc I was developed by International Submarine Technology, Ltd., to search for the remains of the liner *Titanic*. The SeaMarc device is towed in a fish from 100 to 400 meters above the sea floor. It transmits two roughly horizontal beams with a frequency of 27,000 to 30,000 hertz. SeaMarc can map a section of sea floor about five kilometers wide; its scale is thus between that of SeaBeam and that of GLORIA. Combining results from the SeaBeam, GLORIA and SeaMarc systems is beginning to provide a rich picture of the spreading centers. On a recent cruise led by William B. F. Ryan of Lamont-Doherty and J. Paul Fox of the University of Rhode Island the SeaMarc system was employed to map the East Pacific Rise between the Clipperton Fracture Zone and the Orozco Fracture Zone. The structures associated with the creation of oceanic crust could be seen as clearly as if they had been observed from the air.

In the past five years satellite data and sonar recordings have filled more detail in the map of the ocean floor than has been added in any comparable period before. As a result earth scientists now have almost as clear a picture of some parts of the midocean ridge as they do of structures on land. To penetrate below the crust and develop a picture of its composition and structure, however, other techniques are needed.

Much of the current understanding of the layers of the oceanic crust comes from recordings of seismic waves, both those generated naturally by earthquakes and those generated experimentally by explosions or special air guns. Indeed, the definition of the crust was originally formulated on the basis of work with seismic waves. The Mohorovičić discontinuity, or Moho, which divides the crust from the mantle, was first detected because of its capacity to reflect seismic waves.

The speed with which seismic waves travel depends on the temperature, pressure and composition of the rock medium. Hence recordings made at a distance from an earthquake or an experimental explosion can yield clues to the makeup of the intervening material. Two main types of waves are studied in the seismic work: body waves, which tend to travel through a particular layer, and surface waves,

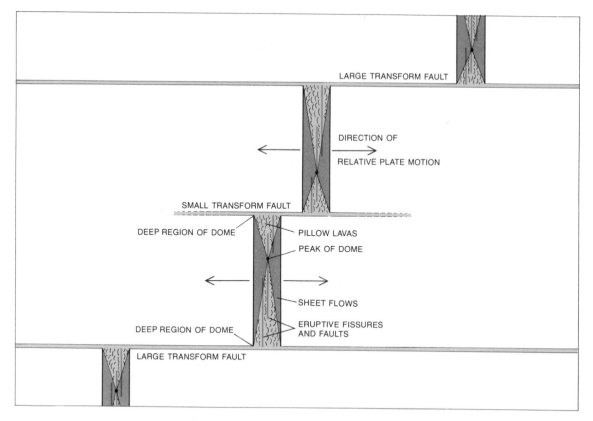

LARGE TRANSFORM FAULT

DIRECTION OF

RELATIVE PLATE MOTION

SMALL TRANSFORM FAULT

DEEP REGION OF DOME

PILLOW LAVAS

PEAK OF DOME

SHEET FLOWS

DEEP REGION OF DOME

ERUPTIVE FISSURES
AND FAULTS

LARGE TRANSFORM FAULT

Figure 2.9 HYPOTHESIS proposed by J. Francheteau and R. D. Ballard to explain how the spreading center operates includes the idea that each large dome between transform faults is an individual spreading cell: an area where the creation of oceanic crust proceeds independently. In this schematic diagram two complete domes are shown. The broken line of eruptive fissures from which lava flows marks the ridge crest. Each dome has a peak between faults; the dome slopes down to deep regions next to the fault. Near the peak the ridge crest is covered mainly by sheet flows. Farther down, the slope is covered by pillow lavas. Which form of lava is predominant depends on the position on the dome. If each dome is a spreading cell, the crust could be created in thin ribbons extending from the sides of the spreading cell.

which tend to travel along the boundary between two layers.

Both body waves and surface waves can consist of two other types of waves: P waves and S waves. P waves are analogous to sound waves in air. As the P wave passes, the rock is compressed and expanded only in the direction of wave motion. S waves subject the rock to shear stresses perpendicular to the direction of wave motion. S waves can travel only through solids; P waves can travel through solids, liquids and gases.

In the late 1950's Russell W. Raitt of the Scripps Institution of Oceanography proposed a seismic model of the crust. In Raitt's model the crust was composed of three layers that could be distinguished on the basis of the P-wave velocity. Layer 1 consisted of the sedimentary cover, Layer 2 the oceanic basement and Layer 3 the oceanic layer. The underlying mantle was designated Layer 4. The P-wave velocity was held to be constant in each layer.

Raitt's hypothesis included three simplifying assumptions. The first was that the wave velocity in the rock always increases with depth. In general this is the case, because the compression of the rock increases its "resonance" as a medium for wave motion. There are, however, exceptions to the rule, and the velocity pattern in the crust is complex.

Raitt's second assumption was that the boundary between any two layers of the crust is a horizontal plane. His third assumption was that the layers are quite thick in relation to the characteristic length of the waves. In general a seismic wave can give information only about features that are considerably larger than the length of the wave. Features that are smaller than the wavelength have little effect on the path of the wave and therefore cannot be detected when the wave motion is analyzed. Experimental explosions (which were the main source of seismic waves in geologic work in the 1950's) generally give rise to waves with a wavelength of from .5 kilometer to two kilometers; thus they yield information only about components of the crust that are a few kilometers thick or thicker.

Recent work has shown that all three assumptions are drastic oversimplifications. Even the idea that the velocity in the layers is constant has now been called into question. Paul Spudich and John A. Orcutt of Scripps and G. Michael Purdy of Woods Hole have put forth a model of a crust in which layers are defined not by the absolute wave velocity but by the velocity gradient: the change in wave velocity with depth. Recent seismic work has been much more concerned with Layer 2 and Layer 3 than with Layer 1, where sediment often gives inconsistent seismic results.

According to Spudich and Orcutt, in Layer 2 the velocity gradient is quite steep. The wave velocity increases by from one to two kilometers per second with each kilometer of depth in the crust. In mathematical notion such a gradient is written as 1 to 2 s^{-1}. In Layer 3 the gradient is about .1 s^{-1}. Thus the P-wave velocity in Layer 3 is almost uniform, which makes Layer 3 the best defined region of the crust from the seismic point of view. In Layer 4, the upper mantle, the wave velocity exceeds 7.8 kilometers per second.

Some seismologists argue that there is an additional low-velocity zone between Layer 3 and Layer 4 corresponding to the transition from the crust to the mantle, but the evidence for such a layer is scanty. It is known that the transition from the crust to the mantle takes place over a distance of three or four kilometers. Waves with a wavelength short enough to give a high-resolution picture of the Moho, however, become quite attenuated in traveling to the bottom of the crust and back to the surface, and therefore little is known in detail about the transition layer.

Purdy took a seismic profile southwest of Bermuda in oceanic crust created during the Mesozoic era about 140 million years ago. Carefully allowing for the effects of a variable sediment cover, he found that the crust was 7.2 kilometers thick above a 500-meter transition zone to the mantle. The P-wave velocity at the top of the crust was five kilometers per second. Layer 2, defined as the region where the gradient was greater than .64 s^{-1}, was 2.3 kilometers thick. Layer 3, with a gradient of .1 s^{-1} or less, was 4.9 kilometers thick.

In the crust where Purdy did his work Layer 3 is subdivided into two levels: an upper level 1.7 kilometers thick with a gradient of .1 s^{-1} and a lower level 3.2 kilometers thick where the waves show no change in velocity with depth. The velocity in the lower level of Layer 3 is seven kilometers per second. Thus the oceanic layer (Layer 3), seismically defined, is about five kilometers thick and the basement layer above it (Layer 2) is half as thick.

Several workers have attempted to go beyond the structural profile given by the velocity gradients to infer the composition of the rock layers. This is generally done by comparing the P-wave velocity with the S-wave velocity for rock at a given depth in the crust. Different kinds of rock have characteristic ratios of P- to S-wave velocity and hence some deductions about the rock type can be made on the basis of the seismic data. Unfortunately for the seismologist the wave-velocity ratios for the rock types are not unique. The velocity change from Layer 2 to Layer 3 could therefore be interpreted as either a change in metamorphic rock type (from greenschist above to amphibolite facies below) or a change in lithologic type (from basalt above to gabbro below).

It is possible that knowledge of the exact composition of the lower layers of the oceanic crust will have to wait until drills penetrate deeper in the crust than they have done so far. In the winter of 1981 a drill on the *Glomar Challenger* in the six-nation International Program for Ocean Drilling succeeded for the first time in going deeper than one kilometer. The drill hole, 1,076 meters deep and designated 504-B, is in the Costa Rica rift, which crosses the East Pacific Rise between the Galápagos Islands and South America (see Figure 2.10).

Although a one-kilometer hole seems modest compared with the full seven-kilometer depth of the crust, the achievement is a considerable one because of the problems of drilling in the ocean. The

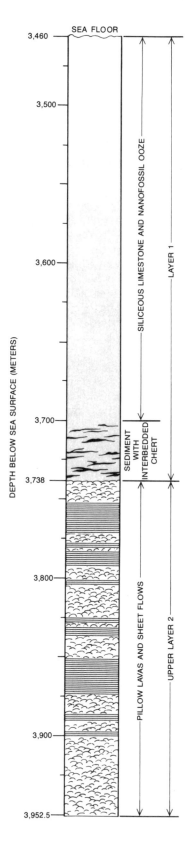

Figure 2.10 DEEPEST HOLE yet drilled in the oceanic crust is Hole 504-*B*, on the Costa Rica rift between the Galápagos Islands and South America. Drill cores from Hole 504-*B*, such as the one shown in schematic form, are providing new information about the structure of the crust. The depth below the sea surface is given at the left; the core depicted extends for about half the full depth of the hole. The crust on the Costa Rica rift is about six million years old. It is covered by a sediment layer 275 meters thick, made up mainly of the remains of microscopic marine plants and animals. Below 275 meters is the oceanic basement, composed of pillow lavas and sheet flows. Intriguingly, it has been shown that the in situ pressure at the bottom of the core in the figure is less than the pressure at the ocean floor. The difference in pressure could pull seawater down into cracks in the crust.

difficulties encountered in drilling holes in the oceanic crust have included the twisting and breaking of the drill string and severe wear on the bits. The drilling of Hole 504-B was relatively free of such problems. What does Hole 504-B tell us about the crust?

The crust at the drill hole is six million years old. Because the Pacific at the latitude of the Costa Rica rift is warm and biologically active, the surface of the crust is already covered with a layer of sediment 275 meters thick. Below the sediments the upper 575 meters of the basement consists of pillow lavas and the rocks called breccias and hyaloclastites, which are produced by the melding of small pieces of fractured basalt into a single mass under pressure. Between 575 and 780 meters the first dikes appear and extensive breccias are interspersed with a few pillow lavas. From 780 meters to the bottom of the hole are massive basalts, abundant dikes and a notable lack of either pillow lavas or fractured material.

Recordings of the in situ pressure deep down in Hole 504-B made by Roger N. Anderson of Lamont-Doherty and Mark Zoback of the U.S. Geological Survey are clarifying the circulation of water through the upper layers of the crust. Anderson and Zoback employed an inflatable device called a packer to isolate a section of the hole. The pressure in the isolated zone was measured from a "go devil" developed by the oil industry to explore oil and gas wells. Intriguingly, Anderson and Zoback observed that the water halfway down the hole is under less pressure than the water at the top. The difference in pressure is about eight bars. (One bar is 14.7 pounds per square inch.)

The mechanism of the remarkable reversal of the pressure gradient is not yet understood. The underpressure in the hole could, however, be due to a heat-convection cycle in the mantle. The relatively low pressure in the lower layers of the crust could help to explain the hydrothermal circulation; it would tend to pull seawater from the ocean floor down into the cracks in the crust.

The depth of the hydrothermal circulation is currently a matter of considerable controversy. It has been suggested that the circulating water penetrates deeply enough to have a role in regulating the operation of the magma chamber under the midocean-ridge axis. If enough seawater reaches the lower layers of the crust, it could cool the magma and cause it to solidify. The lava would therefore stop flowing onto the surface of the ridge until the plates pulled apart enough to renew the decompression of the mantle rock.

For the circulating water to have such an effect it would have to penetrate into the gabbro in Layer 3. Hole 504-B did not go deep enough to settle the question, but it is significant that the rocks recovered from a depth of 600 meters on down to the bottom of the hole show a pattern of alteration that is quite compatible with the fluxing of heated seawater through them. Moreover, the rocks from the deepest part of the hole are among the most extensively altered. Such alteration would have had to take place before the sediments were deposited, capping the basement and preventing the flow of water into it.

The clues to the operation of the magma chamber gained from Hole 504-B are being supplemented by seismic work. In investigating the magma chamber both the reflection and refraction of seismic waves are observed. In the reflection experiments air guns towed behind a ship are generally employed as the energy source. The waves from the air guns pass down to the crust and are reflected upward to the ship, where their travel time and amplitude are recorded. In the refraction experiments the energy propagates at the interface between rock layers and can be recorded some distance away on the sea floor with an ocean-bottom seismometer or at the sea surface. By combining reflection and refraction results the speed of the waves through the structures of the crust can be calculated.

The speed of seismic waves is reduced considerably by rock in the molten state; therefore a low-velocity zone in the crust could correspond to a magma chamber. There is an area under the East Pacific Rise where the refracted waves are attenuated or much slowed; the area is probably a crustal magma chamber. Magma is an efficient reflector of seismic energy, so that in the reflection work the top of the magma chamber comes out as a strong, roughly flat reflecting surface two or three kilometers below the sea floor. In the spreading center on the East Pacific Rise at nine degrees north and also in the Lau Basin near the Fiji Islands the reflection at the top of the chamber is about four kilometers wide (see Figure 2.11).

Refraction experiments done by Brian T. R. Lewis and his colleagues at the University of Washington also make it clear that the magma chamber is quite narrow. Lewis believes it is even less than four

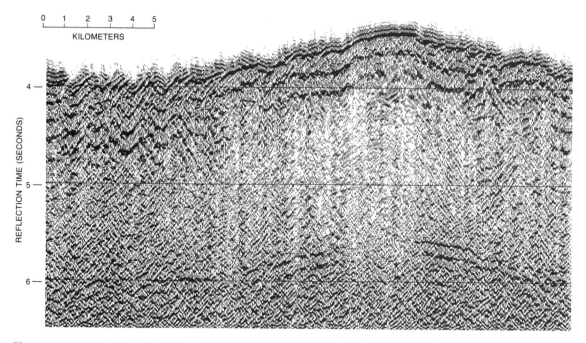

0 1 2 3 4 5
KILOMETERS

REFLECTION TIME (SECONDS)

4 —

5 —

6 —

Figure 2.11 SEISMIC PROFILE at nine degrees 30 minutes north on the East Pacific Rise. The peak on the upper border of the profile is the ridge crest. The dark line at a reflection time of about six seconds is the Mohorovičić discontinuity, or Moho: the boundary between the crust and the mantle. The travel times and amplitudes of the reflected seismic waves yield information about the char- acter of the crustal rock. In a partially liquid magma the seismic waves are considerably retarded. The break in the Moho directly under the ridge crest indicates the presence of magma. The width of the break suggests that at the level of the Moho the magma chamber is less than two kilome- ters across. (Profile by P. Buhl.)

kilometers wide. Thus the entire oceanic crust is created by a thin tube of molten rock running a few kilometers below the crest of the midocean ridge.

The "root" of the midocean ridge, the underlying structures in the mantle, is also being seismic- ally probed. One means of examining the deeper layers is the surface waves called Rayleigh waves. Donald W. Forsyth of Brown University, Nicole Girardin of the Institut de Physique du Globe of the University of Paris and Wolfgang Jacoby of the Uni- versity of Frankfurt studied the passage of Rayleigh waves through young lithosphere in the Pacific and also along the Reykjanes Ridge south of Iceland. They found that the low S-wave velocity extends down to 60 kilometers.

Additional data were supplied by waves from a strong earthquake in Uzbekistan in the southern U.S.S.R. in May, 1976. Rayleigh waves from the earthquake with a period of 300 to 400 seconds, which can make several circuits of the earth, were recorded with ultralong-wavelength seismometers at Los Angeles and at the South Pole. The passage of the long-wavelength waves shows that the low- velocity zone under the East Pacific Rise extends to a depth of 200 kilometers, with the minimum veloc- ity at about 100 kilometers. Thus the midocean ridge has deep roots.

Stuart A. Hall of the University of Houston and his colleagues have advanced a quantitative model of the magma chamber and its root in the mantle. Their work is intended to account for the small gravitational anomalies on the crest of the midocean ridges. As I mentioned above, the ridge is approxi- mately in isostatic equilibrium and so is not asso- ciated with large anomalies in the gravitational field. There are, however, small anomalies directly over the ridge crest. At the Mid-Atlantic Ridge, where the spreading rate is low, there is a negative anomaly; at the East Pacific Rise, where the spreading rate is high, there is a small positive anomaly.

According to Hall and his colleagues, the gravitational anomalies could both be explained by one type of magma chamber. They hypothesize that the density of the material in the magma chamber is about 2.75 grams per cubic centimeter, or 1 percent lower than that of the surrounding rock. The mantle root that is made up of gabbro has a density of three grams per cubic centimeter, or 6 percent lower than that of the surrounding mantle rock. The density of the chamber and the mantle root is thus quite close to the density of the adjacent rock.

Hall and his co-workers conclude that the small gravitational anomalies over the ridges are therefore due not to variations in density but to topographic features at the ridge crest. On the slow-spreading Mid-Atlantic Ridge there is a rift valley along the ridge axis. On the fast-spreading East Pacific Rise the axis is marked by a ridge.

The newest source of information about the structure of the oceanic crust is not seismic waves or gravity data but electromagnetic radiation. It is now possible to measure undersea electric and magnetic fields with considerable accuracy by means of receivers on the ocean bottom. If a source of electric current is put on the sea floor some distance away from such a receiver, the electric field induced by the current travels through the rock. The measured intensity of the electromagnetic indicates how well the intervening section of the crust conducts electricity.

The electrical conductivity of rock is affected by chemical composition, temperature and the extent of melting. A deep electrical conductivity log can therefore be quite informative about the rock of the crust and the upper mantle. As a source of electromagnetic energy Charles S. Cox and Peter Young of Scripps have employed an insulated wire 800 meters long with bared ends. The wire acts as a horizontal electric dipole, with the return flow of current being through the ocean. The wire is laid on the sea floor at the end of a cable attached to a ship. An alternating current with a peak of about 70 amperes is passed through the wire. The energy transmitted by the dipole has a frequency of about one hertz. A pair of cruciform antennas with arms nine meters long are put on the sea floor 19 kilometers away from the transmitter. With this setup Cox and Young have recorded electric field signals as strong as 10^{-10} volt per meter. Since the "noise," or background electric field on the sea floor, is 10^{-12} volt per meter, 100 times weaker than the recorded signal, the finding is significant.

The pattern of received signals shows that there are two layers in the crust with quite different electrical conductivity. The upper layer is at most 1.5 kilometers thick and has a fairly high electrical conductivity: about .1 Siemens per meter. The upper layer corresponds to the relatively young fractured basalts found near the top of Hole 504-B; the measured conductivity in the two places is about equal. In both places the presence of seawater that has penetrated the fractures greatly increases the electrical conductivity.

Below the conductive layer is a region, extending down to about six or seven kilometers, in which the conductivity is much lower: about .004 Siemens per meter. The measured conductivity in the lower region is probably an average for the lower crust and upper mantle. In the deep parts of the crust the conductivity is due to seawater penetrating the gabbro layer. In the mantle the conductivity results from the passage of the electric signal through minerals in the hot rock.

Electromagnetic measurements can yield unique information about the changes in temperature with depth and the presence of molten rock in the deep regions of the crust. Therefore this new method could in the next few years throw much light on the zone near the Moho, which is not easily investigated with other techniques. Novel and intriguing as it is, electromagnetic observation is only one of an array of methods that are currently being harnessed to probe the upper part of the earth under the ocean. Theories about the oceanic crust advance, even more than theories in other areas of science, only in conjunction with new methods of observation. The past decade has been a period of remarkably rapid development of such methods. Given the inevitable lag between observation and the formulation of new theories, in the next decade there could be developed a new and more accurate picture of the thin layer that covers most of the surface of the earth.

The Earth's Hot Spots

These plumes of hot rock welling up from deep in the mantle are a key link in the plate-tectonic cycle. The marks they leave on passing plates include volcanoes, swells and midocean plateaus.

• • •

Gregory E. Vink, W. Jason Morgan and Peter R. Vogt
April, 1985

From deep inside the earth's mantle isolated, slender columns of hot rock rise slowly toward the surface, lifting the crust and forming volcanoes. The plumes well up all over the world, under continents and oceans, both in the center of the mobile plates that make up the earth's outer shell and at the midocean ridges where two plates spread apart. The marks they leave at the surface are superposed on the grand effects of plate motion. Volcanic eruptions and earthquakes associated with plumes occur far from plate boundaries, the site of most such activity; the upwelling currents also form broad anomalous swells in the ocean floor and in continental terrain. These isolated areas of geologic activity are called hot spots.

Mantle plumes are relatively stationary, and so the crustal plates drift over them. Often the passage of a plate over a hot spot results in a trail of identifiable surface features whose linear trend reveals the direction in which the plate is moving. If the plate is oceanic, the hot-spot track may be a continuous volcanic ridge or a chain of volcanic islands and seamounts rising high above the surrounding sea floor. The most prominent example is the Hawaiian Islands; it was a visit there that led J. Tuzo Wilson of the University of Toronto to put forward the concept of hot spots in 1963.

Wilson noticed that to the west of Hawaii the islands disappear into atolls and shoals, indicating they are progressively more eroded and therefore older. The same observation had been made more than a century earlier by the American geologist James Dwight Dana, but Wilson was the first to interpret the age progression as evidence of continental drift. He proposed that the island chain had been formed by the westward motion of a crustal slab over "a jetstream of lava" now situated under Hawaii itself, at the eastern end of the chain. The proposal came at a time when textbooks, including one coauthored just three years earlier by Wilson himself, mentioned continental drift only as an intriguing idea advanced in the 1920's but later discredited.

In the past two decades the idea has become generally accepted as part of the theory of plate tectonics. The earth's crust is now known to be embedded in the rigid plates of the lithosphere, which is between 100 and 150 kilometers thick under continents and about half as thick under oceans; the continual motion of the plates over the partially molten asthenosphere (the portion of the mantle extending to a depth of roughly 200 kilometers) explains the development of ocean basins and the formation of mountain ranges. A major task of contemporary geophysics is to understand how these surface processes are related to the slow convective "creep" of hot rock in the underlying mantle. Hot spots are an important part of this connection.

Indeed, if the upwelling plumes were to stop, the plates would grind to a halt. Ultimately the energy that drives plate motion is the heat released by the decay of radioactive elements deep in the mantle. The plumes provide an efficient way of channeling the heat toward the surface. Their efficiency is attributable to a property of mantle rock: its viscosity, or resistance to flow, is reduced dramatically by relatively small increases in temperature (say 100 degrees Celsius) or in the content of volatile elements such as water. Less viscous material produced by variations in temperature or volatile content tends to collect and rise toward the surface through a few narrow conduits, much as oil in an underground reservoir rises through a few boreholes.

I t would be misleading, however, to say that the plumes propel the plates. Rather, the two are different parts of the same convective cycle. As plates spread apart at a midocean ridge, molten rock from the asthenosphere wells up at the spreading axis to form ocean crust; the new lithosphere cools as it moves away from the ridge and is eventually destroyed at oceanic trenches, where two plates collide and one of them sinks deep into the mantle. The deep mantle feeds the plumes. They in turn empty matter heated by radioactivity into the asthenosphere, which in addition to serving as the source of new sea floor provides a hot and fluid layer for the plates to glide across. The asthenosphere is constantly being destroyed as it cools and attaches to the base of the lithosphere; the boundary between the two layers is essentially a thermal one. Were it not replenished by the plumes, the asthenosphere would soon vanish, and the motion of the plates would stop.

It is worth emphasizing that this "plume model" of the convective circulation in the mantle is just that: a model. The plumes have not been observed directly. The deep mantle can be explored only through the analysis of earthquake waves, and so far the resolution of seismic studies has not been good enough to detect plumes; the upwelling currents may be just a few hundred kilometers in diameter and only moderately different from their surroundings in temperature and density (the properties that determine the seismic-wave velocity in a region).

The indirect evidence for deep-mantle plumes, however, is substantial. Satellite measurements of the earth's gravity field have shown hot spots to be areas of anomalously high gravity and thus of excess mass; the excess mass can be attributed to broad bulges in the surface produced by the upwelling plumes (see Figure 3.1). A second line of evidence comes from geochemical studies of basalts erupted at hot-spot volcanoes. Compared with the basalts dredged from midocean ridges, these rocks are enriched in volatile elements and in other elements such as potassium that are "incompatible" with the crystals of ordinary mantle rock. They also contain anomalous amounts of isotopes derived from radioactive decay processes. The differences in composition suggest that hot-spot lavas are derived from rock welling up from below the asthenosphere, which feeds the oceanic spreading centers. According to the plume model, as material from the deep mantle flows into the asthenosphere, the part rich in volatiles and other incompatible elements melts, and some of it rises to the surface at hot-spot volcanoes.

Recent advances in seismology encourage the hope that someday workers will observe the plumes directly (see Chapter 4, "Seismic Tomography," by Don L. Anderson and Adam M. Dziewonski). In particular, a proposed new global network of seismometers may improve the resolution of seismic studies to the point where it is possible to determine the size of plumes and the depth of their roots.

T he plumes are certainly not uniform; differences in their isotope signatures imply that they come from various depths. Comparisons of the volume and frequency of eruptions at different hot spots indicate they also come in a range of sizes. Furthermore, individual plumes are not immutable. After examining the volume of rock extruded along the Hawaiian hot-spot track, one of us (Vogt) has suggested that the discharge rate of a plume may vary over time. Geochemical evidence supports this conclusion. Jean-Guy E. Schilling of the University of Rhode Island has proposed that plumes consist of rock rising in blobs rather than in a continuous flow. Sometimes a hot spot may fade away entirely,

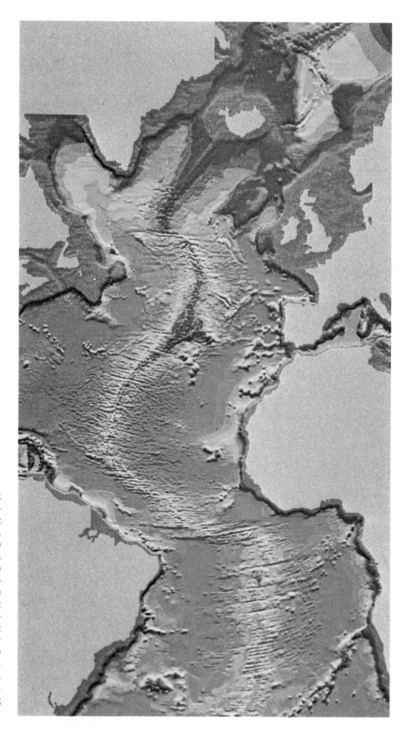

Figure 3.1 MID-ATLANTIC RIDGE in a computer-generated map showing the topographic effect of hot spots. Brown regions are shallow, green regions are deep. Hot spots are present on the ridge axis beneath southeastern Iceland; beneath the Azores, the brown spot between Europe and North America; and in the south Atlantic. Thicker oceanic crust possibly formed by the flow of material along the rift axis away from hot spots. The Iceland hot spot probably also formed the shallow areas to the northwest and southeast during earlier stages of Atlantic ocean opening. (Map by W. F. Haxby of Lamont-Doherty from data compiled by J. E. Gilg and R. Van Wyckhouse.)

and new ones may be formed; from the tracks it appears the typical life span of a plume is on the order of 100 million years. Moreover, the position of a hot spot seems to change slightly. As a result the tracks on the surface are not at all neatly linear as the Hawaiian chain.

Compared with the plates, however, the mantle plumes are relatively stationary. The first evidence of their fixity came in 1970. One of us (Morgan) showed that three volcanic island groups in the Pacific—the Hawaiian Island–Emperor Seamount chain, the Tuamotu Archipelago–Line Island chain and the chain formed by the Austral, Gilbert and Marshall islands—are approximately parallel and could all have been formed by the same motion of the Pacific plate over three fixed hot spots. (see Figure 3.2). In each case the most recent volcanic activity has taken place near the southeastern end of the chain, and the islands and seamounts get progressively older to the northwest. The Pacific plate is currently moving toward the northwest; it switched to that course from a more northerly heading about 40 million years ago. The course change shows up as a bend in the hot-spot tracks.

Because the motion of the hot spots is insignificant, they provide a worldwide reference frame for tracing the absolute motions of the plates with respect to the earth's interior. For some time workers have mapped the paths of the plates in relation to one another and have thereby been able to reconstruct the opening of ocean basins. The boundaries between plates—the ridges and trenches—also move, however, and so the relative motions do not reveal where on the globe a plate was at a given time. Nor do they indicate whether two diverging plates have been moving at the same speed, or whether instead one plate has remained stationary. Such questions can be answered by converting the known relative motions into absolute motions in the hot-spot reference frame, in which each hot spot occupies an unchanging latitude and longitude.

The relative motion of diverging plates—the sea-floor-spreading history—is determined by analyzing magnetic anomalies in the sea floor. Throughout geologic history, at regular intervals averaging about 100,000 years, the earth's magnetic field has reversed its polarity, for reasons that are poorly understood. A record of these reversals is preserved in the oceanic crust. The magnetic minerals in lava erupting from midocean ridges align themselves with the prevailing field, and as the molten rock cools and solidifies, the field direction is permanently locked in the crust.

Figure 3.2 MOTION OF THE PACIFIC PLATE over three fixed mantle plumes has produced three parallel island chains: the Hawaiian Islands and Emperor Seamounts, the Tuamotu and Line islands, and the Austral, Gilbert and Marshall islands. The chains lie in the center of the plate,

The magnetized crust is transported away by the diverging plates in bands that roughly parallel the ridge axis. Each band has a characteristic magnetic anomaly and is made up of crust formed at the same time, and so the bands are called magnetic isochrons. The age of various isochrons, and therefore the sea-floor-spreading rate, has been established through radiometric dating of rocks retrieved in deep-sea drilling expeditions. By superposing

ALEUTIAN TRENCH

HAWAIIAN ISLANDS

HAWAII

LINE ISLANDS

TUAMOTU ARCHIPELAGO

PITCAIRN

EASTER ISLAND

EAST PACIFIC RISE

PERU-CHILE TRENCH

AUSTRAL ISLANDS

proving they were formed by a mechanism different from the one that produced the volcanic island arcs of the western Pacific, which are associated with the subduction of the plate at oceanic trenches. The plumes originate deep in the mantle, and their surface tracks reveal the path of the plates. Active volcanoes, such as Kilauea on Hawaii, are at the southeastern end of the chains. To the northwest the volcanoes are extinct and progressively older.

corresponding isochrons from opposite sides of the spreading axis, one can reconstruct the relative position of the plates at the time the isochron pair was formed. (The superposition in effect removes from the map all sea floor created after the particular magnetic reversal.)

If the motion of one of the plates over the plumes is known, then their relative motion allows the path of other plates in the hot-spot reference frame

to be deduced. The general procedure is to begin with a well-defined hot-spot track on one plate—say a chain of seamounts—and then adjust the more ambiguous tracks until the "best fit" is achieved: the absolute plate motions that best satisfy the constraints established by the hot-spot evidence and the relative motions.

Using this procedure, we have reconstructed the opening of the Atlantic and Indian oceans (see Figure 3.3). The reconstructions can be tested: surface

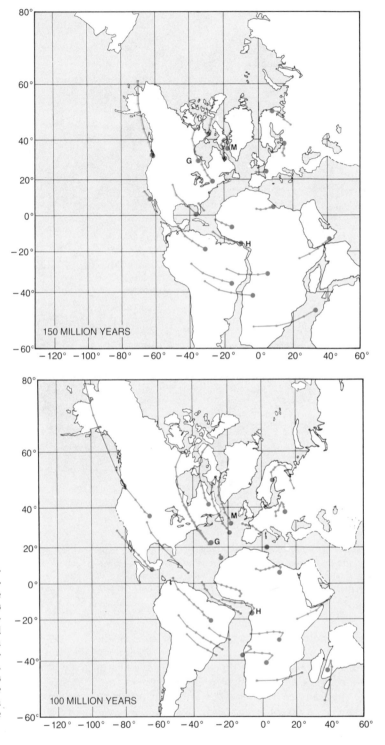

Figure 3.3 HOT-SPOT TRACKS may reveal how the plates have moved with respect to the earth's interior during the opening of the Atlantic Ocean. The tracks consist of extinct volcanoes, magma intrusions and swells in both continental and oceanic crust. Specific plumes discussed in the text are the Great Meteor (*G*), St. Helena (*H*), Madeira (*M*) and Yellowstone (*Y*). Each small dot represents 10 million years of plate motion. Not all hot spots (*large dots*) are present in each reconstruction because new ones form and old ones fade away.

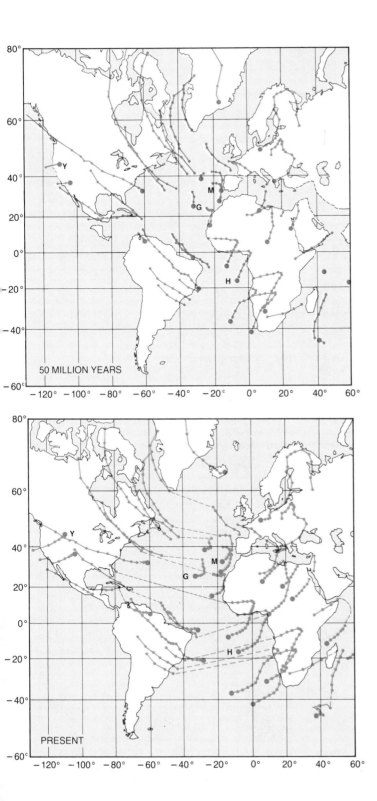

50 MILLION YEARS

PRESENT

features along the hot-spot tracks must by their nature and age fit the hypothesis that they were formed by the passage of a plate over an upwelling plume. This should be true not only along the well-defined portions of the tracks but also in regions where the tracks have simply been extrapolated from the calculated plate motions and evidence of hot-spot activity has not previously been observed.

Although the available data are fragmentary (particularly concerning the ages of sea-floor features), in general the reconstructions pass the test. A good example is the track of the hot spot that formed the Great Meteor Seamount south of the Azores. Two hundred million years ago the area northwest of Hudson Bay on the Arctic Circle was over the Great Meteor plume; 50 million years later the hot spot was under Ontario. The exposure of the Canadian shield from Manitoba to Ontario can be attributed to uplifting of the crust by the plume: in an uplifted area sediment covering the basement rocks is more likely to be eroded away over time.

One hundred million years ago the track had reached the young and narrow Atlantic off Cape Cod. The passage of New Hampshire over the hot spot is recorded by magma intrusions in the metamorphic rock of the White Mountains; the intrusions are between 100 and 124 million years old. For the period from about 100 to 80 million years ago the track follows the trend of the New England Seamounts. Based on radiometric dating of rocks collected from the seamounts, Robert A. Duncan of Oregon State University has shown that the volcanoes get progressively younger toward the southeast along the chain. Their ages coincide with their passage over the hot spot. From the ages and the distances between the seamounts Duncan has calculated the velocity of the North American plate during that period: about 4.7 centimeters per year.

Approximately 80 million years ago the Mid-Atlantic Ridge migrated westward over the plume. The track continues on the African plate and ends at the Great Meteor Seamount. At present the hot spot should be about 500 kilometers southwest of Great Meteor. Although there is a swell in that region of the sea floor, there is no sign of current volcanism; the plume may have become inactive.

Using one or two tracks such as that of the Great Meteor hot spot and the resultant plate motions, one can calculate the tracks of the other hot spots. These must fit the relative plate motions derived from the sea-floor spreading history. If a spreading ridge drifted over a plume, the track continues on the other plate after an interruption caused by the sea floor formed at the ridge as it passed over the hot spot, shown as a broken line in Figure 3.3. The tracks in the figure approximate concentric circles because plate motion is a rotation.

A swell in the ocean floor, like an exposed continental shield, is an area of uplifted crust. Some time ago Robert S. Detrick and S. T. Crough, then at the University of Rhode Island, proposed that a plume produces uplift not by bending the lithosphere but by thinning it, replacing cold, dense lithosphere with hot, buoyant rock from the asthenosphere. After passing over an active hot spot, both sea-floor and continental swells presumably cool and gradually sink back to their former altitude. Swells on the sea floor are interruptions of the process in which the lithosphere cools, thickens and sinks as it moves away from a midocean ridge, eventually plunging into the asthenosphere at a trench.

The hot-spot anomalies, however, are by no means insignificant interruptions. There are some 40 active hot spots, and the swells associated with them have an average diameter of about 1,200 kilometers. Thus swells cover roughly 10 percent of the earth's surface. This observation led Crough and Richard Heestand of Princeton University to suggest that the depth of the sea floor in a particular region is controlled not only by the progressive cooling of the lithosphere but also by the time elapsed since the region passed over a hot spot.

In the same way hot spots could control the thickness of the continental lithosphere. Moreover, the thinning and weakening of continental plates by mantle plumes may produce more dramatic effects than the exposure of basement rock: it may cause them to rift apart. In the early 1970's Kevin C. Burke of the State University of New York at Albany noticed that some hot spots are associated with three-arm rift systems, in which two of the arms have formed a plate boundary whereas the third has failed. The failed rifts form valleys extending into the continents; an example is the Niger River valley.

The reconstructions of the Atlantic opening reveal a number of hot-spot tracks along which continents have subsequently broken up, probably millions of years after the plates passed over the plumes. The track of the hot spot that formed the Madeira Islands, for example, runs between the west coast of Greenland and the east coast of Baffin Island and Labrador; the plume that created St. Helena can be traced along the south coast of West Africa and north coast of Brazil. In the future a rift may develop in the Snake River plain, where the North American plate has been weakened by the track of the hot spot now under Yellowstone National Park.

Mantle plumes explain much of the geologic activity in the center of the plates. As the plates move over the hot spots, however, so do the plate boundaries, including the midocean ridges; unlike the hot spots, the ridges are not anchored deep in the mantle. What happens when a plume is under or near a spreading axis?

A plume directly under a spreading center augments the flow of molten rock welling up from the asthenosphere to form new crust. The crust over the hot spot is therefore thicker than it is along the rest of the ridge, and the result is a plateau rising above the surrounding sea floor. The most striking example is Iceland, a hot-spot island that straddles the Mid-Atlantic Ridge: there the upwelling is so intense and the crust so exceptionally thick that the plateau is above sea level. Geochemically the Icelandic crust is distinctly different from typical oceanic crust; it shows clear evidence of a hot-spot contribution. Gravity measurements indicate that the core of the plume is under the southeastern part of the island. The volcanic peaks there are visible signs of a powerful upwelling current: as much as 5,500 feet high, they are covered by the Vatnajökull glacier. (In 1918 an eruption under the glacier unleashed a flood of meltwater at a discharge rate 20 times that of the Amazon River.)

Some of the material in the strong Iceland plume also seems to spread out under the lithosphere. The lithosphere slopes upward toward a spreading axis, and one of us (Vogt) has proposed that the axis north and south of Iceland has acted as a pipeline, channeling partially molten rock away from the hot spot. In both directions along the ridge the excess plume material produces abnormally elevated topography out to a distance of about 1,500 kilometers. To the south of Iceland the broad plateau tapers to form the typical Mid-Atlantic Ridge. The tapered structure probably arises from the fact that most of the volatile-rich, easily melted plume rock is used up near Iceland. Indeed, Schilling has found that the chemical composition of basalts dredged from the ridge becomes progressively more like "normal" oceanic crust with increasing distance

from Iceland, suggesting that the relative contribution of the hot spot gradually declines.

On the flanks of the ridge south of Iceland there are symmetrical pairs of secondary ridges. Each pair forms a southward-pointed V whose apex is on the spreading axis. These features could have been produced by "waves" of intensified flux or of unusually hot and buoyant material from the plume (see Figure 3.4). A wave traveling down the ridge would generate anomalously thick crust, affecting the area nearest the hot spot first. The elevated crust would then be carried away on each side of the axis by the spreading plates, forming the V-shaped secondary ridges. From the known spreading rate and the angle between the secondary ridges and the spreading axis one can estimate the speed of the plume material; it seems to flow down the axis at a rate of five to 20 centimeters per year.

Because the midocean ridges move, a hot spot is unlikely to be situated under a spreading center for more than a geologically brief period. It is conceivable, however, that a plume might feed a spreading axis from a distance, provided it is close enough to the region in which the base of the lithosphere slopes up toward the axis. This concept helps to explain some unusual surface features in the Iceland area.

The plateau that includes Iceland stretches from Greenland in the west to the Faeroe Islands in the east (see Figure 3.5). The section of the plateau east of Iceland and east of the current spreading center has long puzzled geologists. Its linear trend suggests a hot-spot origin. Yet it could not simply have been formed by the motion of a plate over a fixed plume, because it does not coincide with the track of the Iceland hot spot, which is known from the reconstructions of the early Atlantic. Some workers have interpreted this as a sign that the hot spot has not remained stationary but has instead wandered about, forming the plateau by occasionally punching through the plate. The argument implies that the reconstructions are inaccurate: if plumes are not fixed, they provide no absolute reference frame for mapping plate motions over the mantle.

Our own hypothesis is that the Iceland hot spot has remained stationary and that the Iceland-Faeroe plateau section was made by rock flowing eastward from the hot spot to a now extinct spreading center. The hypothesis can be tested. Presumably the plume would feed the closest point on the ridge. Thus at any time during the formation of the pla-

Figure 3.4 HOT SPOT MAY FEED A RIDGE from a distance, thickening the crust and forming an oceanic plateau. Early in the opening of the ocean basin (*top*) the hot spot is under a thick continental plate moving to the northwest; material from the plume cannot yet reach the spreading center. Millions of years later (*middle*) the motion of the plates has brought the ridge closer and has carried continental shelf over the hot spot. Plume material has begun to flow along the lithosphere to the nearest section of the rise. As the excess material erupts it is carried away on the plates; the V shape of the resulting plateau reflects both the spreading of the plates away from the ridge and their motion with respect to the hot spot. A change in plate motion (*bottom*) forms a bend in the hot-spot track and in the plateau.

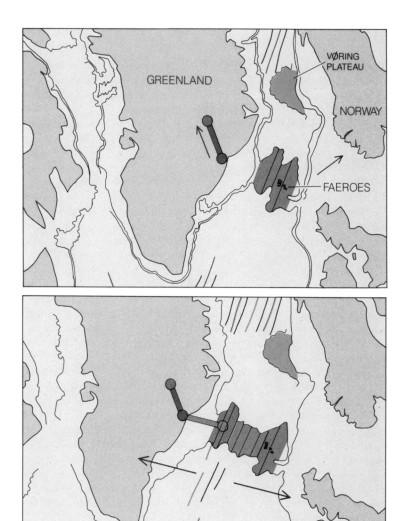

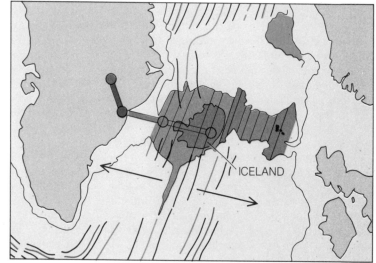

Figure 3.5 GREENLAND-FAEROE
PLATEAU might have been formed by
the Iceland hot spot. The parallel colored lines are magnetic isochrons used
to reconstruct past positions of the
plates. Fifty million years ago (*top*) the
hot spot was under the coast of Greenland and began feeding the ridge. The
V shape of the plateau reflects the hot-
spot track. By 36 million years ago
(*middle*) the plates had changed course;
the change is reflected in the new section of the plateau. At about that time
the spreading axis moved west over the
hot spot, which by then was under oceanic lithosphere. Hot-spot-fed spreading has continued in the west until
now (*bottom*), forming Iceland. At an
earlier time when it was close to a
northern ridge the plume may have
built the Vøring Plateau.

teau a line representing the shortest distance from the plume to the ridge should intersect the center of the plateau. The plateau would be symmetrical about the ridge axis, but not necessarily perpendicular to it. With respect to the hot spot, the plates might have a component of motion parallel to the axis, and the orientation of the plateau would be obtained by adding that component to the relative motion of the plates (perpendicular to the axis). Finally, the age of the plateau at any point would be the same as that of the surrounding sea floor, because the two were formed at the same time. None of these predictions would hold if the plateau were formed by a wandering hot spot that was not feeding a ridge.

To test the model one of us (Vink) reconstructed the opening of the Norwegian-Greenland Sea and the formation of the plateau. The method is the same as that used for reconstructing the early Atlantic: superposing magnetic isochrons reveals the relative position of the plates at the time of a given magnetic anomaly, and the hot-spot track shows the plate motions in the hot-spot reference frame.

During the early opening of the basin, some 50 to 60 million years ago, the Iceland hot spot was under eastern Greenland. Its southerly track reflects the northward motion of the Greenland plate. The passage of the plate over the plume probably produced the extensive igneous rock formations southwest of Scoresby Sound, which from radiometric evidence are judged to be roughly 55 million years old. About 50 million years ago the Greenland continental shelf moved over the hot spot. At that time excess plume material could have begun flowing along the base of the oceanic lithosphere to the spreading center, and the plateau would have started to form. The Faeroe Islands, now at the eastern end of the plateau, would have been created first; their basalts are between 50 and 60 million years old. In the reconstruction of the period the nascent plateau is roughly symmetrical about the spreading axis, and the V shape of its northern edge reflects the northerly motion of the plates with respect to the hot spot.

By 36 million years ago the plates had switched to a more westerly course, causing the hot-spot track to bend to the east. The change is apparent in the geometry of the plateau: the V is split by a younger segment with an east-west heading, perpendicular to the spreading axis. The plateau remains symmetrical about the axis, and a line from the hot-spot position intersects the axis at the center of the pla-

teau. Both observations indicate the plume was continuing to channel molten rock to the ridge.

By that time the hot spot was under oceanic lithosphere, which is somewhat thinner than continental lithosphere. The plume would have thinned it further. Our model assumes that the ridge subsequently jumped to the area of weakened lithosphere, leaving an extinct spreading center on the eastern section of the plateau. Although the existence of such a relic is still being debated, geologic activity seems indeed to have ceased in the east at about the time the spreading axis would have jumped to the west; rocks collected from a drill hole near the center of the eastern section are roughly 40 to 43 million years old.

Sea-floor spreading continued at the western end of the plateau. With the hot spot positioned under the spreading axis, plume material began to flow down the axis, giving the ridge its present tapered structure to the south. The westward-moving plates soon pushed the axis off the hot spot, but the plume continued to feed the ridge. The oldest outcrops on Iceland are found near the east and west coasts, as one would expect on an island formed at a spreading axis; their ages suggest the island was born between 16 and 12 million years ago. Iceland remains geologically active. In the past few million years eastward movements of the spreading axis have once again placed the ridge over the hot spot.

The reconstructions show that a fixed Iceland hot spot could well have produced the observed geometry of the Greenland-Faeroe Plateau. It may also have formed the Vøring Plateau, even though the latter is now 500 kilometers to the north of Iceland. The hypothesis rests on the assumption that a plume will always feed the closest section of a spreading axis. Just before the formation of the Greenland-Faeroe Plateau, when the hot spot was still under Greenland, it may have been closest to a northern ridge segment. During that period it could have produced the Vøring Plateau. The northerly motion of the Greenland plate later brought the southern spreading axis closer to the hot spot, and so the plume switched targets.

Like plate tectonics itself the notion of hot spots is a simple but powerful concept. It explains many features of the earth's surface that once seemed disparate, and further research will undoubtedly lead to the attribution of other effects to upwelling plumes in the mantle. At the same time the concept is appealingly intuitive. Indeed, it is only embellishing the truth a little to suggest the Hawaiians recognized the track of their hot spot centuries before it caught the attention of modern geologists. According to Hawaiian legend, Pele, the fiery-eyed goddess of volcanoes, originally lived on Kauai, at the western end of the island chain. When the god of the sea evicted her, she fled to Oahu. Forced again to flee, she continued to move east, to Maui, and finally to the island of Hawaii. She now seethes in the crater at Kilauea.

Seismic Tomography

By analyzing many earthquake waves with this technique, it is now possible to map the earth's mantle in three dimensions. The maps throw light on the convective flow that propels the crustal plates.

. . .

Don L. Anderson and Adam M. Dziewonski
October, 1984

The outer layer of the earth, the lithosphere, consists of a dozen rigid, crust-bearing plates that ride on the underlying mantle, rearranging continents, forming mountains, creating and destroying oceans. What drives this constant remodeling? Ultimately the motive force is the convective circulation of the mantle. The mantle is solid rock but so hot that over geologic time it deforms easily and flows. New lithosphere is formed at midocean ridges, where hot magma from the mantle wells up between diverging plates. The new surface material spreads outward from the ridges and eventually plunges back into the mantle at oceanic trenches, where two plates collide. Although this model is widely accepted, the origins of the upwelling material and the fate of the subducted plates—in general, the details of the flow in the mantle—have remained unknown, beyond the reach of conventional geophysics and its analytic methods.

In the past few years a new analytic technique called seismic tomography has become available that promises to profoundly enhance knowledge of the earth's internal structure, including the pattern of flow in the mantle. Like its medical analogue, computer-aided tomography or CAT scanning, seismic tomography combines information from a large number of crisscrossing waves to construct three-dimensional images of the medium the rays have traversed. In the case of CAT scanning the medium is the human body and the source of energy is an X-ray generator. The best source of information on the earth's interior is the seismic waves triggered by earthquakes, because they are attenuated only slightly by their passage through the earth; an earthquake of moderate strength radiates waves that are recorded by seismometers all over the world.

Seismic investigations over the past 70 years have revealed much of the earth's average radial structure: the fact that it has a crust, an upper and a lower mantle and an outer and an inner core. Tomography is now adding considerable detail to this simple model. By determining how fundamental properties such as temperature and density vary with latitude and longitude as well as with depth, it is providing the first three-dimensional view of the mantle.

To understand how seismic tomography works one must first understand a few facts about earthquake waves and how they behave. The earth transmits seismic disturbances because it is an elastic medium: it resists deformation, and when a part of it is strained — compressed in volume or sheared in shape — a restoring force acts to return that part to its original condition. A seismic wave is just a traveling strain triggered by the release of stress by an earthquake. Seismic waves travel fastest through regions of the earth that are most resistant to deformation.

Every earthquake radiates waves that plunge through the body of the earth as well as ones that travel along the surface. The body waves fall into two categories: compressional waves and shear waves. Compressional waves are like sound waves in that they consist of periodic compression and dilation of the rock along the direction of travel. Because earthquakes are precipitated by slippage or shearing across a fault, however, they also radiate shear waves, which are analogous to electromagnetic radiation in that the direction of vibration is transverse, or perpendicular, to the direction of propagation. Like electromagnetic waves, shear waves can be polarized, in which case they vibrate in a single transverse direction.

Surface waves also are of two principal types. Rayleigh waves, the first type, have both a compressional and a shear component; they cause rock particles to move elliptically in the vertical plane that lies between the earthquake focus and the detector. In contrast, Love waves are polarized shear waves vibrating in the horizontal plane parallel to the earth's surface. Although both Rayleigh and Love waves travel along great-circle paths at the surface, they extend deep into the mantle and thus provide information on its structure.

The velocity of shear waves is a function of the medium's rigidity, which is a measure of its resistance to shearing stress. Liquids, for instance, are not rigid and hence do not transmit shear waves; that is how it was discovered that the earth's outer core is liquid. In the case of compressional body waves and the compressional component of Rayleigh waves, velocity is a function of both rigidity and another elastic property of the medium: its incompressibility. (Similarly, sound, a compressional disturbance, travels faster through water than through air, which is more compressible, and faster still through ice, which is both incompressible and rigid.)

Cold material in general tends to be more rigid and more incompressible than hot material, and seismic waves therefore traverse cold regions of the earth's interior more rapidly. Hot material in turn has relatively low density and is generally associated with ascending flow in the mantle, whereas cold matter sinks because it is denser than its surroundings. Wave velocity also depends on a small-scale property of the medium: the orientation of crystals in the material. Mineral crystals of the kind that make up the mantle have three axes, and each axis exhibits a different degree of stiffness. The most rigid axis is "fast." If over a large region the fast crystallographic axes become aligned, say, by a current in the mantle, waves whose polarization (direction of vibration) or direction of propagation is parallel to the fast axes will accelerate in that region.

The velocity of seismic waves thus contains indirect information on the pattern of flow in the mantle. Extracting that information, however, is not easy. The velocity of a single ray, as computed from its arrival time at a seismic station, is simply an average over the ray's entire path and does not reveal where the wave has been slowed or speeded up. Moreover, the average is generally over a great distance, because large expanses of the earth, particularly the oceans, are without seismic stations, and because earthquakes tend to occur only at plate boundaries. To draw inferences about such properties as density and temperature in the earth's interior it is necessary to combine information from many rays; the more the better.

Because of a vast expansion at the seismic data base, it is now possible to construct much more detailed images of the earth's seismic-velocity structure through the application of tomography. Recent improvements in the data base have been twofold. For the past seven years a global network of sensitive digital seismometers has been recording long-period surface waves in computer-readable form on magnetic tape. At the same time the International Seismological Centre (ISC) near London has been collecting reports from more than 1,000 ordinary seismic stations around the world. Most of these stations are relatively insensitive to surface waves, but they record body waves, which have a much shorter period, from earthquakes all over the world. As a by-product of its effort to pinpoint the location of some 10,000 earthquakes per year, the ISC has collected arrival times for several million seismic rays.

Although medical tomographers have an advantage over seismologists in that they can control the source of the radiation as well as the detection apparatus, the analytic technique is fundamentally the same. In CAT scanning, X rays are used to map density variations in the human body and thereby reveal internal organs and other structures. The absorption of X rays is greatest in dense regions of the body, and so these regions show up on the image as shadows. On a conventional radiographic image it is often difficult to distinguish overlapping structures, particularly when they have similar densities; CAT scanning overcomes this problem by mathematically recombining information taken from many X rays sent through the body along different paths. The result is horizontal slices that when stacked show internal structure in three dimensions.

In tomographic studies of the earth's interior it is the velocity of seismic waves and not their absorption that is measured, and the resulting images are maps of "slow" and "fast" regions in the mantle. These anomalies are found, just as in CAT scanning, by combining the information from many crisscrossing rays (see Figure 4.1). If the velocity of an individual seismic ray deviates from the expected value (taken from seismological tables that list average travel times for seismic waves according to the surface distance between the epicenter and the seismometer), the anomalous mantle mass that caused the deviation could lie anywhere along the ray's path. If, however, another ray crosses the first one at some point, then the measured velocity of the second ray provides a constraint on the velocity of the first ray at the point where they intersect. A dense mesh of many intersecting rays creates a network of mutual constraints that make it possible to map the velocity structure of the region covered by the mesh. The denser the mesh is, the greater are the resolution and the accuracy of the map.

In practice finding velocity values that satisfy all the constraints requires a complex mathematical procedure and a large computer to carry it out. Essentially the technique consists in solving a set of simultaneous equations for each unit region of the mantle. On the right side of the equations are the known travel times of all the rays that traverse the region. On the left side of each equation is a series of terms with associated velocity parameters. Solving the problem means finding the values of the velocity parameters that produce the closest fit between the expansions of terms and the observed travel times. This is an example of the "inverse problem" in seismology: working backward from observations to a model of the earth's structure.

The body-wave arrival times collected by the ISC and the surface-wave data recorded by the network of long-period seismometers are complementary sources of information on the mantle. Body waves are the only direct way of probing the lower mantle, which extends from the bottom of the upper mantle at 670 kilometers to the boundaries of the core, which lies at 2,900 kilometers. One of us (Dziewonski) has recently derived, from a data base of more than 500,000 rays, a model of the lower mantle that resolves structural features with horizontal dimensions of 2,000 to 3,000 kilometers and a vertical extent of 500 kilometers (see Figure 4.2). Robert W. Clayton and his colleagues at the California Institute of Technology Seismological Laboratory have derived a more detailed model using a larger set of data and special techniques for enhancing resolution. The large-scale features of both studies are similar; for example, island arcs such as the Philippines, Japan and Indonesia show up at hot regions underlain by cold material in the upper mantle. This pattern reflects the cold descending plates of the subduction zones overlain by hot mantle material. In addition, both studies have detected large, fast anomalies near the core-mantle boundary at 2,900 kilometers.

A body wave that plunges into the earth from an earthquake focus and later reemerges at a seismometer follows a relatively short, nearly vertical path through the upper mantle, and with the current distribution of seismic stations it is impossible to achieve a sufficiently dense mesh of body-wave paths to image that layer. On the other hand, the entire surface of the earth is fairly well sampled by Rayleigh and Love waves that "see" deep into the upper mantle. Long-period, or low-frequency, surface waves sample the mantle to greater depths than waves with relatively short periods, just as a long swell in the ocean reaches farther below the surface than short chop. The new long-period digital seismometers register surface waves whose velocity is affected by structure down to a depth of about 700 kilometers, slightly below the boundary between the upper and lower mantle.

With our colleagues Ichiro Nakanishi, Henri-Claude Nataf and Toshira Tanimoto at Caltech and John H. Woodhouse at Harvard University we have analyzed these surface-wave data and have constructed tomographic images showing the lateral and vertical extent of velocity anomalies in the upper mantle. (see Figure 4.3). So far we have calculated, for each of some 60 large earthquakes that occurred between 1977 and 1982 and that provide the best available geographic coverage, the velocity of surface waves of about six frequencies traveling

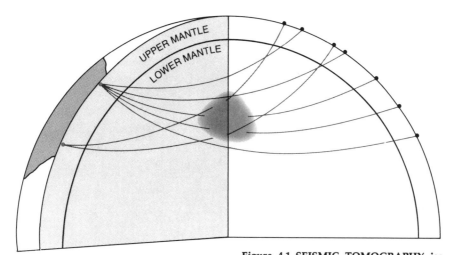

Figure 4.1 SEISMIC TOMOGRAPHY isolates velocity anomalies (*color*) in the mantle by combining information from many seismic waves traveling from earthquakes (*colored dots*) to seismic stations (*black dots*) along crisscrossing paths. Waves that miss the anomaly show the normal travel times for their surface distances. Waves that penetrate the anomaly are slowed down or speeded up. A dense mesh of intersecting rays can be used to define the anomaly and to measure the velocity with which it transmits waves; the structure of the anomaly must account for the observed deviations in travel time of all the rays that traverse it. The lower mantle is mapped with body waves (*top*) that dive into the earth. Long-period surface waves (*bottom*) that "see" to great depth offer the best coverage of the upper mantle.

from the event to each digital seismometer. The different frequencies give information on the velocity structure of the mantle within different, overlapping depth intervals. On the average we use data from about 20 seismic stations; because surface waves travel from earthquake to station along both the short arc and the long arc of a great circle, we obtain, for each quake and each frequency, measurements of average velocity along about 40 different paths. To prepare a map of the mantle at a particular depth or a vertical cross section along a particular great circle we typically combine between 400 and 1,000 different paths, converting the averages along many arcs into specific velocities in each region.

Plate-tectonic theory gives us some idea of what to expect in mapping the upper mantle. Under midocean ridges, volcanic regions and regions such as the Red Sea where rifts are forming in continental plates, seismic velocities should be slow; these are areas where hot, less dense mantle is welling up toward the surface, melting as it rises. Below stable continental "shields," where plates have remained at the surface for billions of years and so have had ample time to cool, one would expect seismic waves to travel anomalously fast, at least in the shallow mantle. At greater depths fast anomalies should be found in regions that have been cooled by the subduction of oceanic lithosphere, which has itself cooled while at the surface.

To a certain extent our results confirm these expectations. At a depth of 150 kilometers we have observed slow seismic velocities under most of the world's tectonic and volcanic regions, including the midocean ridges. In contrast, the Canadian, Brazil-

Figure 4.2 EARTH MODEL derived from seismic tomography showing lateral variation in seismic velocity. White lines at 670 kilometers and 2,900 kilometers correspond to rapid changes in seismic velocity with depth. The vertical cross sections through the entire mantle are along three great circles: along the Equator and from Hudson Bay through central Asia and through the western Pacific. Red and yellow indicate hot, upwelling regions, where seismic velocity is anomalously slow; blue regions are cold, dense, and fast, and green indicates average velocity for a particular depth.

ian, Siberian, African and Australian shields are all fast. The velocity differences, in fact, are greater than can be explained by temperature alone; there must also be some as yet undetermined lateral variations at this depth in mineral composition or in the extent of melting of the mantle.

The fact that ridges and volcanoes are underlain by hot mantle at shallow depths is not surprising. The more interesting question, much debated in geophysical circles, concerns the depth to which these hot thermal anomalies extend. Our maps suggest they reach at least to the 400-kilometer level, but the surface expressions are often offset by large distances from the mantle source.

At a depth of 350 kilometers, for example, the globe-girdling midocean ridge system is no longer continuous but is broken up into isolated segments. The central Mid-Atlantic and the Southeast Indian Ocean ridges are actually underlain by fast material, and the slow anomaly associated with the East Pacific Rise has either vanished or is laterally offset. At 550 kilometers there is even less relation between mantle and surface features: most of the Atlantic and Indian Ocean ridges are fast, whereas the Siberian shield, cold and therefore fast near the surface, is slow. Clearly the midocean ridge system is not simply the surface expression of vertical upwelling currents. Instead it seems to be fed by lateral transport of hot material from a few broad thermal anomalies in the upper mantle.

The tomographic maps also confirm some general predictions, based on plate tectonics, about the location of subducted lithosphere. Since the breakup of the supercontinent Pangaea some 200 million years ago and the subsequent formation of the Atlantic Ocean, the Pacific has been shrinking (see Figure 4.4). North and South America have been encroaching on it from the east, and in the west Pacific lithosphere is being overridden by eastern Asia, Australia and various large islands. In mapping the velocity structure of the mantle at a depth of 350 kilometers we found fast regions under eastern Asia, South America and the central and southern Atlantic—right where one might expect to find cold, subducted Pacific lithosphere. At 550 kilometers the fast anomalies are larger and even more prominent; the one in the west includes most of the western Pacific and all of Australia.

Interestingly, seismic velocities in the western Pacific at shallower depths—200 kilometers and less—are slow. This too is expected from plate tectonics. When cold oceanic lithosphere enters the mantle at a subduction zone, it displaces hot mantle upward. At the same time volatile compounds (mainly water) in the subducted sediment and crust lower the melting point of the mantle, and friction melts part of the mantle "wedge" above the descending slab. Hot, buoyant magma rises to the surface and forms volcanoes. The island arcs of the western Pacific, including Japan and the Philippines, are volcanic archipelagoes situated above subduction zones.

Our results show that mapping seismic-velocity anomalies is an accurate way of locating hot and cold regions and hence of locating ascending and descending currents in the mantle. By themselves, however, the maps offer little information on the final leg of the convective circulation: the horizontal flow of subducted material from trench to ridge. Fortunately clues to the pattern of this horizontal flow can be extracted from seismic data by taking into account the fact that the velocity of seismic waves in the mantle also depends on the direction in which the waves travel through a given region. As a result of differences in crystal orientation waves propagate faster in a particular azimuth, or horizontal direction, than along other azimuths. This property of the mantle is called azimuthal anisotropy.

Much of the shallow mantle (the layer above a depth of 400 kilometers) is composed of olivine, a magnesium-iron silicate. Olivine crystals are anisotropic with respect to seismic waves: one axis of the crystals is significantly faster than the other two axes. If a mass of crystals were oriented randomly, the effects of their anisotropy would cancel, but field studies have shown that olivine crystals tend to be aligned by flow in the mantle. Over very large areas their fast axes are oriented parallel to the flow direction; this can be thought of as analogous to the magnetization of a piece of iron by an external magnetic field.

By mapping the fast directions of seismic waves it is thus possible to get an idea of the horizontal flow in the mantle. Tanimoto and one of us (Anderson) have prepared a map showing the fast directions of Rayleigh waves, which are most sensitive to physical properties in the 200-to-400-kilometer depth interval (see Figure 4.5). Because Rayleigh waves travel fast in each direction along the fast crystallographic axis, the flow directions on the map have a 180-degree ambiguity, but by assuming that the flow is generally from subduction zone to ridge the

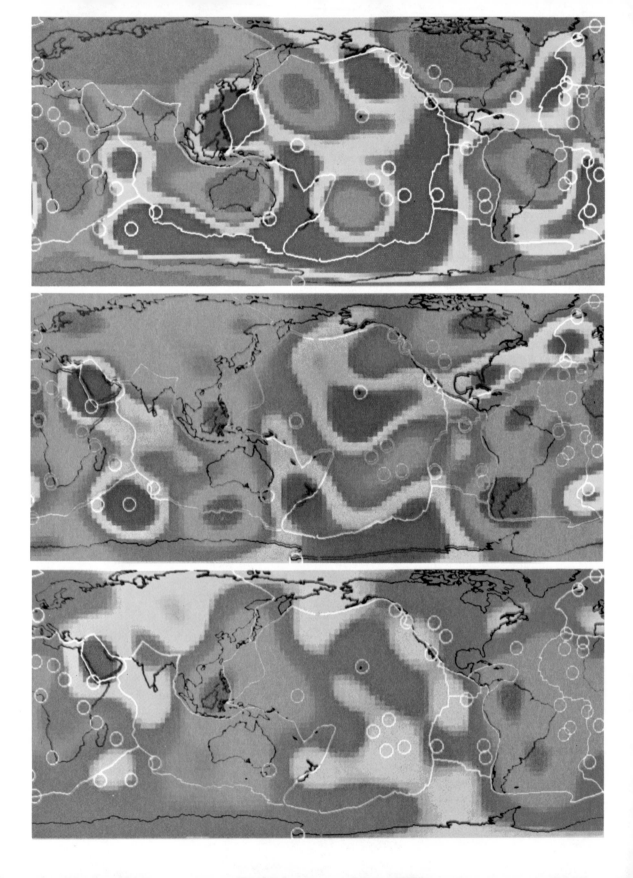

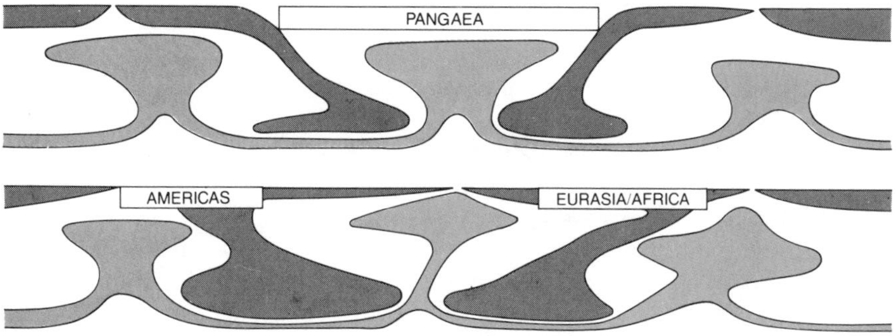

Figure 4.4 BREAKUP OF PANGAEA some 200 million years ago opened the Atlantic and has led to the steady shrinking of the Pacific. The process is illustrated schematically with vertical cross sections of the upper mantle before (top) and after (bottom) the breakup. The hot anomalies (red) are stable features; the one under Pangaea became the source of the Mid-Atlantic Ridge. As the Americas, Australia and Asia override Pacific lithosphere (gray), however, the subduction of surface material cools the mantle and distorts the hot upward flow. The subducted lithosphere creates fast anomalies that make the continents appear to have deep, cold "roots."

ambiguity can usually be resolved. When they are combined with the seismic velocity maps, the maps of azimuthal anisotropy provide powerful new constraints on theoretical models of convection in the mantle. They show, for example, that the upper mantle under central North America is flowing from north to south, whereas under Siberia the flow is probably in the opposite direction.

By exploiting the anisotropy of the mantle in another way, we can distinguish on a single map between regions of horizontal flow and regions of vertical flow. Rayleigh waves vibrate in the vertical plane, whereas Love waves vibrate horizontally. In a region in which the crystals are aligned the velocities of a Rayleigh wave and of a Love wave traveling along the same path will thus be different. In general the horizontally polarized shear-wave velocity of Love waves should be greater in regions of horizontal flow, because the fast axes of the olivine will lie in the flow plane and in the direction of vibration of at least some of the Love waves passing through the region. Conversely, the Rayleigh waves' vertically polarized shear velocity should be greater in ascending or descending currents.

The data from the "polarization anisotropy" with that of seismic velocity combine to indicate that the hot regions beneath midoceanic ridges and continental rifts are regions where mantle material is flowing vertically upward. Likewise, the data suggest the presence of downward motion at subduction zones in the western Pacific. The data also suggest considerable evidence of large-scale horizontal transport of both hot and cold material.

Figure 4.3 TOMOGRAPHIC MAPS show shear velocity of surface waves in the upper mantle. At 150 kilometers (top) surface tectonic features are still evident: midocean ridges in the Atlantic, the eastern Pacific off South America and the southern Indian Ocean have slow velocity (red). At 350 kilometers (middle) there is less correlation between seismic velocity and surface features, but cold subducted Pacific lithosphere shows up as fast regions (blue) under the western Pacific and South America; these fast anomalies are even more pronounced at 550 kilometers (bottom). The darkest colors indicate seismic velocities that differ by 2 percent from the average (green) for that depth. White lines and circles represent plate boundaries and surface hot spots. (Maps based on a study by J. H. Woodhouse and A. M. Dziewonski.)

Until the advent of seismic tomography geophysicists had no direct way of mapping convection in the mantle. From data collected by seismometer networks in earthquake-prone regions they were able to determine seismic velocity, temperature and density as global averages at various depths, and in a few regions it was even possible to discern broad lateral variations in these properties. But the data and the analytic technique needed to construct global maps of lateral variations were lacking.

In the absence of direct observational evidence workers inferred what they could about mantle

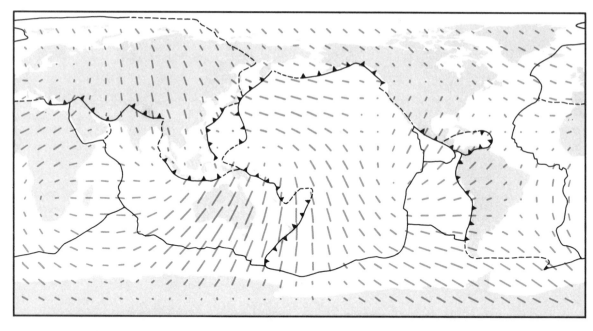

Figure 4.5 HORIZONTAL FLOW in the upper mantle can be mapped by determining the fast propagation direction of Rayleigh waves. Such waves travel fastest either way along the fast crystallographic axes, which tend to be aligned by flow; the 180-degree ambiguity in the flow lines is partially resolved by assuming that the flow is generally from subduction zones (*toothed black lines*) back to ridges (*solid black lines*). The map was prepared at Caltech using 200-second waves, which are sensitive to structure at depths of 200 to 400 kilometers.

flow from indirect evidence gathered at the earth's surface: mainly variations in gravity and in surface elevation (or ocean depth). In general hot upwelling regions are expected to exhibit anomalously high gravity and high elevation. The problem, however, is that gravity and elevation are integrated functions of the way density varies with depth. In other words, a given gravity or elevation anomaly could be produced by an anomalous mass at any depth. As a result variations in these quantities cannot serve to locate unambiguously the density anomalies that drive convection in the mantle, and so they are a poor guide to the nature of the convective flow.

Geophysicists have also placed heavy emphasis on numerical models of simplified convective systems. One significant simplification in most models has been the assumption that the convective flow is two-dimensional, with material moving in a vertical plane and from ridge to trench at the surface and from trench back to ridge in the mantle. Most numerical calculations have also assumed that viscosity, or resistance to flow, is uniform throughout the

mantle. Such assumptions were adopted in part because more complicated flow models would require calculations that strain the power of even the largest computers.

Not surprisingly, reality as it is now being revealed by direct seismic mapping of the mantle turns out to be a good deal more complex than the simple models. For one thing, it is three-dimensional: both cold and hot material flow in many lateral directions, and the spreading at the surface and the return flow in the mantle are by no means in the same vertical plane. Thermal anomalies under midocean ridges, under continental rifts and even under volcanic regions can be traced to great depth, but they are offset from the surface features and are not simply vertical sheets of rising magma. The thermal anomalies and the density variations they induce control the motions of the plates to some extent, but plate tectonics in turn affects the location of the anomalies: the subduction of cold lithosphere cools the mantle, and thick continental lithosphere at the surface insulates the mantle below and causes it to heat up.

Furthermore, in most areas of the world there is, contrary to the assumption of many numerical models, a low-viscosity layer in the upper mantle under the lithosphere. This weak layer partially decouples the plate from the mantle. As a result the moving plates drag mantle along with them and distort the underlying circulation. Some years ago Bradford H. Hager, now at Caltech, and Richard J. O'Connell of Harvard did flow calculations taking the weak layer into account; their results, unlike those of simpler models, are generally consistent with our map of horizontal flow based on zaimuthal anisotropy.

Seismic tomography allows a reversal of the previous sequence of analysis: instead of trying to infer the existence of density anomalies in the mantle from the earth's gravity field, investigators can use seismic maps of the density distribution to explain observed variations in gravity. Drawing on the results of two independent tomographic studies, one by Clayton and Robert Comer of Caltech and the other by one of us (Dziewonski), Hager and his

colleagues have shown that the large-scale variations in the gravity field—broad gravity highs over the central Pacific and Africa and lows over the Indian Ocean and Antarctica—can be traced to large density anomalies in the lower mantle. Smaller-scale fluctuations in gravity, on the other hand, seem to be at least in part the result of density variations in the upper mantle. Our maps show that the gravity highs in the North Atlantic between Iceland and the Azores, in the South Atlantic centered on the island of Tristan de Cunha and in the southwestern Indian Ocean between Madagascar and Kerguelen are all underlain by slow regions in the mantle, corresponding to hot, upwelling plumes, at depths of 200 to 400 kilometers; a hot plume in the upper mantle also appears to contribute to the gravity anomaly in the central Pacific that Hager has linked to the lower mantle. Although the hot material has relatively low density, these plumes are nonetheless expected to produce gravity highs because they increase surface elevation and thus raise the center of gravity under a particular surface area.

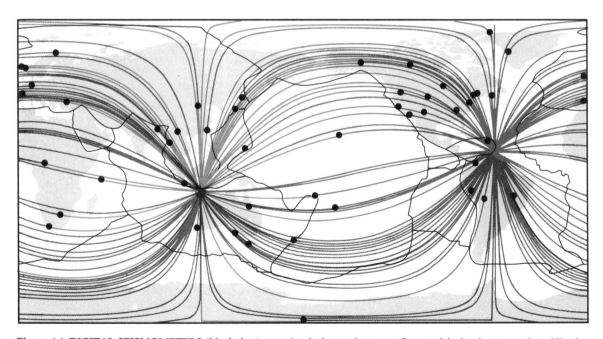

Figure 4.6 DIGITAL SEISMOMETERS (*black dots*) record long-period surface waves that travel from an earthquake (*colored dot*) along both the minor arc (*color*) and the major arc (*gray*) of a great circle. Installing new stations, particu- larly on the ocean floor and in land areas such as Siberia that are not near earthquake zones, would tighten the mesh of intersecting rays and improve the resolution of tomography.

The flow model Hager used to make his calculations assumes that subducted lithosphere sometimes penetrates the seismic boundary between the upper and the lower mantle. The question of whether material can flow across the 670-kilometer boundary, or whether instead the convective circulation in the mantle is divided into two separate layers, is a central issue in geophysics. Tomography has not yet settled the debate. Our maps show apparent continuities across the boundary in some regions, such as the anomalously hot central Pacific, but discontinuities in others. Moreover, the continuities could be explained without assuming that material is transferred across the boundary: the lower mantle may simply heat regions of the upper mantle as a stove heats a pot, by conduction. Unfortunately the boundary is in a depth interval in which the resolution of both surface-wave and body-wave tomography is poor. The solution is to do a tomographic analysis combining both types of data, and in particular to analyze body waves whose paths are confined to the upper mantle because they are reflected or refracted off the 670-kilometer boundary and other discontinuities.

Although the results we have obtained are a striking improvement over conventional seismology, the sparseness of the present network of digital seismic stations limits the resolving power of surface-wave tomography, even in the shallow mantle (see Figure 4.6). Only velocity anomalies that are either very large, about 2,000 kilometers in radius, or very pronounced can be mapped. Recently seismologists from nearly 50 universities formed the Incorporated Research Institutions for Seismology (IRIS), a nonprofit corporation with the objective of modernizing seismological research facilities. Among other projects IRIS plans to improve both the quality and the quantity of the digital data by increasing the number of long-period seismic stations to at least 100 and by operating about 1,000 portable seismometers. A detailed study of a subduction zone using the portable seismometers in conjunction with the global array of permanent digital seismometers might determine, for example, whether descending plates do indeed plunge into the lower mantle. Together the new permanent and portable seismic stations should produce noticeably sharper global images of the mantle and of the hidden flow that shapes the surface of the earth.

SECTION

. . .

PLATES IN ACTION

Introduction

In the conventional plate tectonic model rigid plates were expected to have little tectonic complexity at divergent an transform plate boundaries. In this simplistic view, as new plates are created at divergent boundaries, they rapidly (geologically speaking) migrate away from the boundary into a midplate position. Transform boundaries involved horizontal movement with little or no deformation of the plates.

By contrast considerable faulting and folding has long been thought to characterize subduction zone or consuming margins. As one plate descends beneath another, material is detached from it and accumulates against the edge of the overriding plate in a region of complex folding and faulting known as an accretionary wedge or prism (see Figure 1.6). As the plate reaches depths greater than 100 kilometers, parts of it melt to form magmas that rise to high in the crust, or even to the surface, to form igneous rocks, such as granites and high strato-volcanoes such as Mount Etna, Mount Saint Helens and Mount Fujiyama. Some rocks of the accretionary prism or the crust over the descending slab recrystallize to form metamorphic rocks that are unique to these environments. Indeed, as mentioned in "Chapter 1, 'The Continental Crust,'" many workers have suggested that much of the material that makes up continents was added to them at consuming margins.

Recent work at constructive and transform margins has demonstrated, however, that they also contain many more complexities than postulated in the conventional rigid plate view.

Chapter 5, "How Continents Break Up," outlines some recent advances in our understanding of how continents rift apart. Among other things, it describes propagating rifts recently discovered along several midoceanic spreading centers. In conventional plate tectonics, the ridge segments separated by transform segments, such as those illustrated in Figures 2.5 and 2.10, remain apart the same length and distance. Motion along the transform fault occurs only between the active ridge segments. In contrast, propagating rifts are ones that lengthen at the expense of their counterpart on the other end of the transform fault. The dying ridge branch shortens in time as the plates spread. The result is a curious V-shaped pattern in the magnetic anomaly patterns called a pseudofault.

Propagating rifts are useful in consideration of continental breakup because they explain several puzzling features. One is that when continents around the Atlantic are reattached in a predrift configuration, they overlap in some places. In addition, the oldest oceanic crust adjacent to continents is not all the same age; it gets younger in one direction. For example, in the south Atlantic Ocean, the oldest crust is younger in the north than it is in the south. In the Gulf of Aden, the oldest oceanic crust is younger to the west. These observations suggest that when continents rift, they part like a zipper, rather than separating simultaneously all along their length.

Chapter 6, "Oceanic Fracture Zones," outlines recent discoveries on these zones, which consist of active transform faults that offset segments of the midoceanic ridges as well as inactive fracture zones beyond the active ridge crests. As mentioned above, transform faults are one of the three principal types of plate boundaries, where the relative motion of the plate is parallel to the boundary.

As plates move, they rotate about an imaginary point on the earth's surface, called the pole of rotation. Transform faults are supposed to be segments of small circles on the earth's surface that are concentric about the pole of rotation. In principle they should be of very small width.

The transform faults of theory turn out to be oversimplifications of transform faults in fact. Transform fault zones contain some of the sharpest relief of the oceans. They generally do not exactly describe small circles as the theory holds they should. In addition many are not narrow single zones, but are wide belts of disturbance tens or even hundreds of kilometers in width. Rocks exposed in fracture zones represent samples from a cross section of the entire oceanic crust and even from the mantle below.

How Continents Break Up

They rift apart over millions of years, and in the process they deform. A study of the rifting is beginning to reveal the properties of the plates that compose the earth's crust.

. . .

Vincent Courtillot and Gregory E. Vink
July, 1983

The theory of plate tectonics holds that the crust of the earth consists of plates some 100 kilometers thick that are moving with respect to one another. It accounts for a wealth of facts. The distribution of earthquakes and volcanoes around the world is largely explained by the motions of the plates: their creation at midocean ridges; their destruction by collision or at subduction zones, where they plunge downward into the earth's mantle; their friction as they slide past each other along what are called transform faults. Moreover, the similarity in shape between the edges of two continents that today are separated by thousands of kilometers (for example the eastern edge of South America and the western edge of Africa) shows that they rifted apart. In other words, they were once on a single plate. Corroborating evidence is supplied by the distribution of plant and animal fossils on the continents.

If it is assumed that plates do not deform as they move, their relative motions can be calculated by means of mathematical theorems describing the motion of a rigid body on a sphere. Indeed, the motions of most of the earth's large plates over the past 150 to 200 million years have been worked out with considerable precision. On the other hand, the motions of plates that are assumed to be rigid cannot yield more than an understanding of their kinematics. The assumptions that plates are rigid means it must also be assumed that their boundaries have a fixed geometry, and this latter assumption makes it impossible to propose a realistic model of how continental crust breaks up. For example, the breakup that created South America and Africa must be taken to have happened instantaneously along a line, a most unlikely event.

Actually there is ample evidence that continental crust does deform. Mountain ranges, which often result from the collision of plates, are the most obvious evidence. In addition the distribution of earthquakes at plate boundaries suggests that the boundaries are not narrow, and the occurrence of earthquakes well away from plate boundaries suggests the release of stress, and thus the occurrence of deformation, in the middle of a plate. The work we shall describe in this chapter suggests that the breakup of continental crust must be seen as a process taking millions of years along a zone several hundreds of kilometers wide. It suggests, moreover, that the crust at such a zone is unlikely to behave like anything rigid.

The key evidence supporting the theory of plate tectonics is found on the ocean floor in the form of magnetic anomalies (see Figure 5.1). Fundamentally the volcanic material that rises into place to become new oceanic crust at the axis of a midocean ridge is hot enough for the minute magnetic domains in it to become aligned with the magnetic field of the earth. As the material cools, locking the alignment into place, it spreads away from the ridge axis. Behind it along the axis newer crust is rising into place. Now, the magnetic field of the earth varies in an irregular pattern, and its polarity reverses approximately five times every million years. Therefore as the new oceanic crust spreads away from the ridge axis two divergent magnetic "tapes" record the pattern of reversals. This makes it possible to measure the age of the ocean floor and also determine the velocity at which plates have diverged. That velocity proves to vary from less than a centimeter per year to more than 15 centimeters. Moreover, when the ridge axis is offset by a series of transform faults, the magnetic pattern in the oceanic crust should reproduce the offset.

There are instances, however, where the magnetic pattern fails to match the offset. These mismatches were initially explained in terms of sudden jumps of the ridge to a new location. The problem is that some of the jumps are so numerous and so closely spaced in time that the process is best seen as the continuous evolution of the boundary of a plate. Consider the peculiar V-shaped magnetic patterns found in the Pacific, first near the Galápagos Islands and then on the Juan de Fuca Ridge off the Pacific Coast of North America. A study of those patterns led Richard N. Hey, who was then working at the Hawaii Institute of Geophysics, and his colleagues (particularly Frederick K. Duennebier of the Hawaii Institute, W. Jason Morgan of Princeton University and Peter R. Vogt of the U.S. Naval Research Laboratory) to develop a model in which a rift propagates continuously through the oceanic crust at a rate comparable to the rate at which plates spread apart at the rift.

As Hey and Tanya M. Atwater of the University of California at Santa Barbara explain it, the process occurs where two rifts are offset by a transform fault. In the process is modeled in steps, each step is a sudden jump in which the tip of the rift propagates forward, establishing a new fault and moving oceanic crust from one side of the rift (that is, from one plate) to the other (see Figure 5.2). The result is a V-shaped wake consisting in part of small pieces of crust bounded by abandoned faults. If the process is taken to be continuous, as it in fact is in the crust of the earth, the wake is a V-shaped discontinuity in the pattern of magnetic anomalies. Hey pointed out that the discontinuity can mistakenly be attributed to a fracture. (The attribution is mistaken because no slippage between plates ever occurred along the discontinuity.) He called each limb of the V a pseudofault.

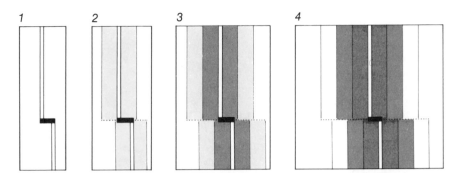

Figure 5.1 MAGNETIC ANOMALIES on the ocean floor are the strongest evidence that new oceanic crust spreads away from each midocean rift. The rift itself (*1*) is offset by a transform fault (*black*). New crust arises continuously. At first it is hot; thus its intrinsic magnetization becomes aligned with that of the earth. As it cools and solidifies, the magnetization is locked into place; hence at each side of the rift a band of magnetized crust diverges (*2*). The earth's field occasionally reverses its polarity. Each such reversal entails two new magnetic anomalies (*3, 4*).

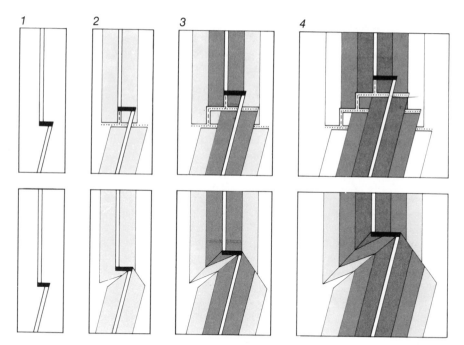

Figure 5.2 DISRUPTED PATTERN of magnetic anomalies is evidence that rifts propagate. In the top row a model proposed by Richard N. Hey and colleagues is shown in steps. Two rigts at an angle to each other are offset by a transform fault (*1*). New oceanic crust rises into place and spreads; then the tip of the southern rift propagates northward, transferring a block of the new oceanic crust from east to west of the rift, that is, from one plate to another (*2*). Successive episodes of sea-floor spreading and rift propagation (*3*, *4*) produce a V-shaped zone of magnetic discontinuity. In the bottom row of the illustration the process is continuous; thus the sea floor spreads while the rift is propagating.

The breakup of continental crust must also involve the propagation of rifts, yet the details of the process have been elusive. In the early 1970's J. Tuzo Wilson of the University of Toronto and Kevin C. Burke and John F. Dewey of the State University of New York at Albany suggested that there might be a link between "hot spots," triple junctions and continental breakup. They suggested that if a continent becomes stationary with respect to the underlying mantle, thermal anomalies (hot spots) in the mantle might make the continent dome upward and rupture in a three-branched pattern of rifts. (Doming and rupture often seem to give rise to a three-branched pattern. They do it, for example, in the crust of a pie baking in the oven.) The three rifts might then develop into two active rifts and an aborted one. (Burke and Dewey thought they could see examples along the rifted margins of the Atlantic.) The active rifts propagating away from a number of junctions could eventually join up and thereby fragment the continent.

More recently two other proposals have been made. First, Morgan has proposed that rifting often occurs along a hot-spot track. Here the motion of a continent over a thermal anomaly in the underlying mantle gives rise to a line of volcanoes at the surface. Under the surface, meanwhile, the thermal anomaly thins and weakens the crust, making it prone to rift. Second, Hey and Vogt have suggested that a process resembling the model of a rift propagating through oceanic crust might account for some of the rifting on the continents. On the ocean floor, however, the model requires that one rift recede while another one advances. Thus the model can only modify the geometry of preexisting plate boundaries. It cannot apply to the breakup of a continent, a process in which a new plate boundary arises.

At this point magnetic anomalies are again important evidence. In 1977 one of us (Courtillot), working with Jean-Louis Le Mouël and Armand Galdeano of the University of Paris, made a magnetic map of part of the western Gulf of Aden and part of the adjacent land mass, the Afar depression in Dijibouti and Ethiopia (see Figure 5.3). Afar is a natural laboratory where sea-floor spreading can be studied without the hindrance of thousands of meters of overlying water. Like Iceland, it is an exposed part of the worldwide midocean-ridge system. Many investigators believe much can be learned about the fundamental process of rifting in these dry-land settings. As early as 1938 the French paleontologist and philosopher Pierre Teilhard de Chardin was convinced that Afar was the place where Alfred Wegener's theory of continental drift should be tested. More recently the investigators who dived in research submarines to the Mid-Atlantic Ridge at 37 degrees north latitude as part of the French-American Mid-Ocean Underwater Survey (FAMOUS) were struck by the resemblance of the rift at Afar to what they had seen under almost three kilometers of water.

The clearest clues to what has happened at Afar are the magnetic maps (see Figure 5.4). The maps reveal that two types of crust are juxtaposed. In one type the magnetic anomalies are intense and form parallel bands. The pattern is typical of the one that results from sea-floor spreading. The bands can be correlated with the periods between reversals of the earth's magnetic field, and so they can be dated. In the other type of crust the magnetic anomalies are wider, rounder and less intense. They describe a magnetic quiet zone, or MQZ.

The MQZ forms a huge reclining V that abruptly interrupts the oceanic pattern. It therefore interrupts the bands of magnetic anomalies that mark sea-floor spreading; hence the edges of the MQZ can be dated rather precisely. Between 45 and 46 degrees east longitude in the Gulf of Aden the edges are about 10 million years old; they intersect the magnetic anomaly representing ocean floor of that age. Toward the west the edges become progressively younger. Near Lake Asal in Afar they are very young indeed. A number of earthquakes and the birth of a volcano near Lake Asal in November 1978, confirm that the vicinity of the lake is a site of tectonic activity today.

The simplest interpretation of these various observations is that a rift is propagating westward into preexisting oceanic and continental crust The edges of the MQZ were created by past propagation, which formed new oceanic crust in its wake. The Gulf of Aden is the result. The opening angle of the

V formed by the edges is about 30 degrees. It arises by simple trigonometry from the rate of the sea-floor spreading that has opened the Gulf of Aden (1.5 centimeters per year) and the rate at which the tip of the rift has moved westward (three centimeters per year).

Support for this interpretation has come from a study of seismic activity in the region undertaken by Jean-Claude Ruegg and Jean-Claude Lépine of the University of Paris. They observe that the foci of most earthquakes under the Gulf of Aden cluster in a band approximately 10 kilometers wide that closely follows the rift. (The position of the rift is deduced from a mapping of the bottom of the gulf and from the position of the youngest magnetic anomaly.) The band of earthquake foci extends from the Gulf of Aden into the Gulf of Tadjurah, a beak-shaped westward extension of the Gulf of Aden. After that the band coincides with the Ghoubbet-Asal Rift, which runs from the Gulf of Tadjurah to Lake Asal. Apparently it ends near Lake Asal, where, we suggest, the tip of the propagating rift is currently situated. West of Lake Asal the seismicity becomes diffuse, suggesting that there the crust is deforming over a wider area (see Figures 5.5 and 5.6).

Further evidence comes from the age and the chemical composition of the basalts at Afar, as they have been determined by Michel Treuil of the University of Paris, Jacques Varet of the Bureau de Recherches Géologiques et Minières and Olivier Richard of the University of Paris at Orsay. Basalts are the volcanic rock that composes oceanic crust. The basalts found on the edges of the Gulf of Tadjurah have a tholeiitic composition: they are poor in the mineral olivine but contain the mineral orthopyroxene. Such basalts are formed by the large-scale partial melting of the mantle under the mid-

Figure 5.3 AFAR DEPRESSION in northeastern Africa is one of two places on the earth (the other is Iceland) where a midocean ridge comes up onto dry land. The axis of each such ridge is a rift where two great plates diverge and new oceanic crust rises into place. The positions of the rifts at Afar are controversial. According to the authors, one rift courses westward from the Gulf of Aden into the Gulf of Tadjurah, the beak-shaped body of water extending into the image from the right. Its tip, currently at Lake Asal just west of the Gulf of Tadjurah, is propagating into the depression and will ultimately join with a rift whose tip is moving southward in the Red Sea (*upper right*). The rifting will then have given rise to a new ocean basin.

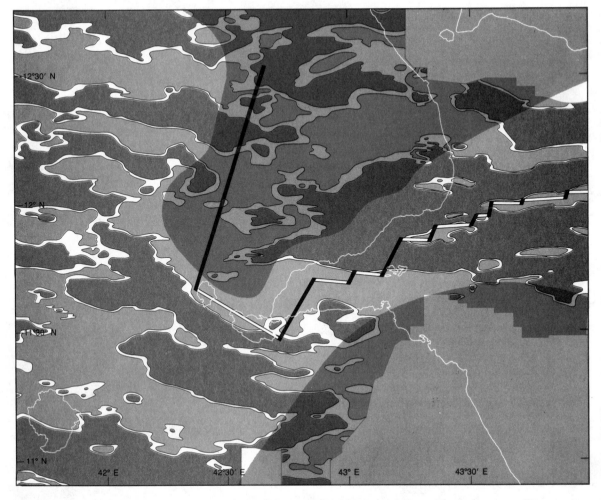

Figure 5.4 MAP OF MAGNETIC ANOMALIES at Afar and on the floor of the Gulf of Aden reveals that different types of crust are juxtaposed. In the gulf the anomalies (shown in alternating blue and purple) are intense and elongated, in a pattern typical of oceanic crust. Elsewhere, and particularly toward the northeast, the anomalies are subdued. They mark out a funnel-shaped region called the magnetic quiet zone, or MQZ (*gray*). The MQZ may represent continental crust deformed by the propagation of the rift along the floor of the gulf.

ocean ridge. Closer to Lake Asal and the tip of the rift the basalts are more alkaline, that is, they are richer in sodium and potassium. This indicates a lesser degree of partial melting than tholeiites undergo. The age of the basalts at each location closely matches the age deduced for the oceanic crust at that location from the measured magnetic anomalies.

In sum, the principal sign of a propagating rift at Afar is a distinctly oceanic pattern of magnetic anomalies that is disrupted by the intersection of

the pattern with the rifted edge of a continent. The magnetic quiet zone at Afar is therefore best taken to be continental crust that has deformed in the course of the propagation.

Signs that rifts have propagated elsewhere in the world, so that continents deformed, emerge from reconstructions: the efforts made by earth scientists to determine how continental fragments fitted together before they rifted apart. The usual method is to try to match up their edges. Thus each fragment is rotated about an axis passing through the center of the earth until magnetic anomalies of the same

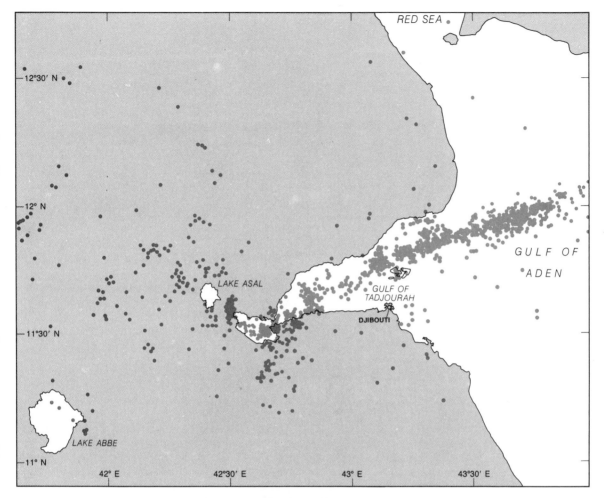

Figure 5.5 FOCI OF EARTHQUAKES at Afar were recorded between 1974 and 1980 by investigators from the Institut de Physique du Globe in Paris. They confirm the position of the rift. Scattered foci west of the tip of the rift suggest that the crust there is now deforming.

age are made to coincide. If the edge of each fragment arose at a single instant in time, so that the edge is precisely parallel to the oldest magnetic anomalies on the ocean floor that has arisen between the fragments, the reconstruction will succeed: it will match the fragments precisely and at the same time yield the age of the rifting. If, however, the rift has propagated, the reconstruction will not quite succeed: the edge of each fragment will not have arisen all at once, and so the edge will cut across magnetic anomalies.

In such instances bringing together the edges of two continents will not reproduce the configuration they had before they rifted apart. A more accurate reconstruction will match the points where rifting began. The continents will then overlap in places, but that is quite appropriate. The overlaps represent the stretching of the continental crust that took place as the crust was broken apart by propagating rifts.

This appears to be what happened in the Gulf of Aden. If, for example, an attempt is made to reconstruct the relative positions of the African plate and the Arabian plate 10 millions years (the age of what is called magnetic anomaly 5), it is found that the gulf closes up (see Figure 5.7 and 5.8). Moreover, it is found that the plates begin to overlap at 43 degree 30 minutes east longitude. Ten million years ago the

Figure 5.6 VIEW DOWN THE AXIS of the rift at the Gulf of Aden is afforded to the east from a point on the western shore of Lake Asal, 155 meters below sea level. The rift is at the center. It is flanked on each side by faults running parallel to the rift that produce a set of escarpments. The pattern is much like the one observed at midocean ridges from research submarines under almost three kilometers of water.

tip of the rift was there. This explains the observation, made by A. S. Laughton and his colleagues at the National Institute of Oceanography in England in 1970 and confirmed recently by James R. Cochran of the Lamont-Doherty Geological Observatory, that anomaly 5 and anomalies older than anomaly 5 cannot be found west of 43 degrees 30 minutes east longitude.

Overlaps also emerge in reconstructions of the North Atlantic made by one of us (Vink). Specifically overlaps between northern Greenland and Spitsbergen suggest that the edge of each of these bodies of land stretched by as much as 100 kilometers. Still another example emerges from the fit of South America and Africa. The classic reconstruction was done by Sir Edward Bullard of the University of Cambridge. Bullard used the 1,000-meter-depth contour lines in the Atlantic to represent the edge of each continent. He rotated the lines together. A computer program found the fit that minimized gaps and overlaps. Even so, there remained a gap of 250 kilometers between the southern tip of Africa and the edge of an undersea plateau that includes the Falkland Islands. The Deep-Sea Drilling Project of the Scripps Institution of Oceanography has since shown that the Falkland plateau is continental crust. It is a submerged part of South America. The gap cannot have existed. When it is closed in a reconstruction, there are overlaps between South America and Africa.

In any case magnetic surveys of the South Atlantic show that the oldest magnetic anomalies on the ocean floor do not parallel the edges of South America and Africa. The surveys confirm that propagating rifts created the South Atlantic. The oldest magnetic anomalies (one with an age of 135 million years near the edge of Africa and one with an age of 127 million years near the edge of South America) are found only in the southernmost part of the Atlantic. The floor of the Atlantic was therefore spreading in the south at a time when Africa and South America were still joined in the north; the rift had not yet propagated that far.

The available evidence suggests the following picture of how a rift propagates to break up a continent and form a new ocean basin. First a system of fractures develops in a plate, possibly along a preexisting weak zone such as an old suture zone (the locus of an earlier collision between continents), a large strike-slip fault system (a fault system where the slippage is horizontal) or, as Morgan has suggested, a hot-spot track. Let us assume for the sake of simplicity that the fractures lie in a linear zone, and let the average width of the zone be 200

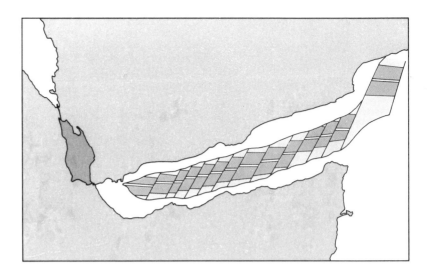

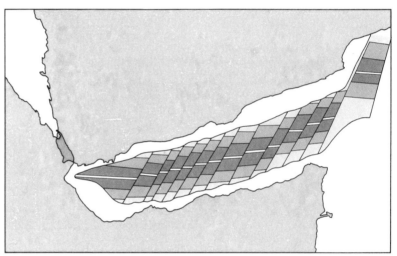

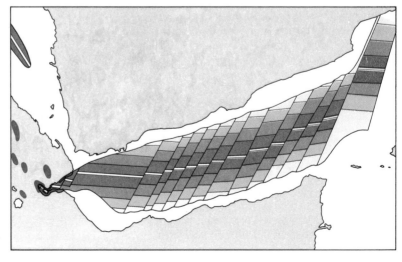

Figure 5.7 RECONSTRUCTIONS OF THE GULF OF ADEN as it looked 10 million years ago (*top*) and seven million years ago (*middle*) are made by taking the present-day map of magnetic anomalies on the floor of the gulf (*bottom*) and removing the anomalies that are younger than those ages. The direction of rift propagation (east to west) is plain. Overlaps between Arabia and Africa (*dark gray*) show that the plates stretched as they rifted. In the present-day map a number of active volcanic ranges (*solid color*) indicate deformation ahead of the tip of the rift. The Red Sea rift is at the upper left of the present-day map. It too is flanked by magnetic anomalies.

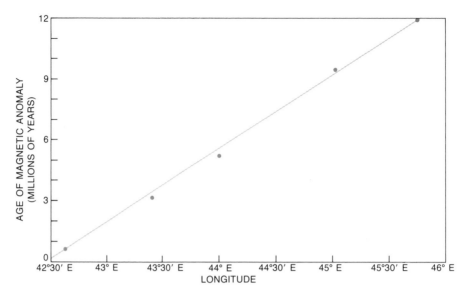

Figure 5.8 SPEED OF PROPAGATION of the rift in the floor of the Gulf of Aden is deduced from the reconstructions. Each point plots the westernmost limit of a magnetic anomaly (*horizontal axis*) and the age of the anomaly (*vertical axis*). The slope of the line connecting the points implies a velocity of three centimeters per year. The end of the line at an age of zero and a longitude of about 42 degrees 30 minutes demonstrates that the tip of the rift is now at Lake Asal.

kilometers, as it is in the Rift Valley of East Africa. As the fractures develop, the continental crust is stretched and thinned. Nevertheless, the relative motion of the plates that will emerge from the process is negligible: the total stretching is limited to a few tens of kilometers and occurs over periods of a few tens of millions of years.

As the stretching progresses, increasing lengths of the future plate boundary have decreasing strength. The mechanics of the process are unlikely, however, to be uniform. Thus the plates remain attached at a number of "locked zones" (see Figure 5.9). These zones cannot prevent the failure of the crust that lies between them. Hence basaltic oceanic crust begins to rise into place. In short, rifting begins. Each rift elongates rapidly through the thinned crust. Then the tips of the rifts enter the remained locked zones and begin to break them apart.

Let us now consider a single locked zone. As in Hey's model of sea-floor spreading, a sequence of discrete steps is easier to explain than the actual continuous process. In the first step the tip of a rift has reached a locked zone from each side. Along the length of each rift new oceanic crust is being created. It fills the space freed as the two plates move apart. Inside the locked zone, however, no new crust develops. Therefore the locked zone must stretch.

In the next step the tips of the rifts penetrate an incremental distance into the locked zone. In this increment the crust of the locked zone stops stretching; the creation of oceanic crust takes over. Hence an increment of the locked zone is carried away by each plate. The part of the locked zone not yet invaded by the rifts continues to stretch. The process goes on until the rifts meet and the two plates separate completely. What remains of the locked zone is a deformation in the edge of each newly formed continent. The deformation is maximal at the point where the continents last touched.

The model can be refined so that the process is continuous. It then emerges, naturally enough, that if parameters such as the rate at which the rifts propagate change with time, the deformation of the edge of the continent produces various shapes. For instance, the shape of certain overlaps that arise in attempts to match the edges of continents that broke apart suggests that the propagating rifts may slow down as they first cut through a locked zone

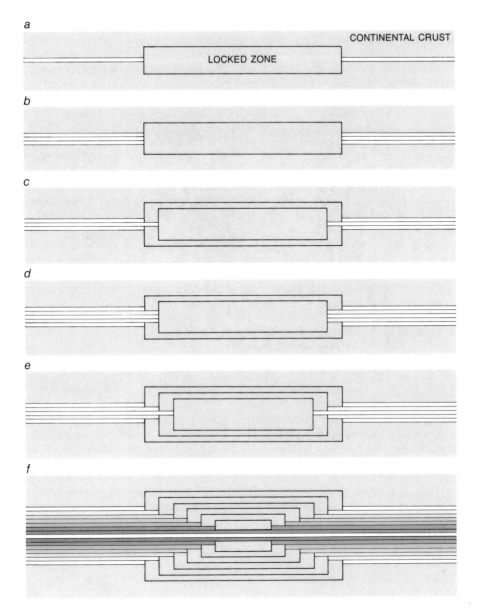

Figure 5.9 RIFTING OF A CONTINENT into two separate plates according to a model developed by V. Courtillot. The process begins along a preexisting zone of weakness. The crust, however, is heterogeneous and so a number of areas resist. One such "locked zone" is shown; it prevents two rifts from meeting (*a*). Along the length of each rift new oceanic crust rises and fills the space freed as the incipient plates move apart. The locked zone must stretch (*b*). The tips of the rifts then begin to invade the zone (*c*). Still more oceanic crust is created and what remains of the locked zone stretches again (*d*). The tips of the rifts move closer together (*e*). Ultimately the rifts meet and the plates separate (*f*). The locked zone has become a deformation in the edge of each new continent.

and then may accelerate in the final phase of the rupture. (The result in such a case is a deformed zone with a flattened tip.) More generally the model predicts that the strain in a continental plate that breaks up occurs in a band some 100 kilometers wide at the margin of each newly formed plate and also in the locked zones, which stretch dramatically. The disruption of the crust in a locked zone may be responsible for the lack of intense magnetic anomalies in the magnetic quiet zone at Afar.

Yet if Afar is to be viewed as a locked zone in the act of deforming, a second propagating rift in addition to the one that opened the Gulf of Aden must be identified. The East African Rift Valley, which has a long history of very small motion, may be an aborting rift and is not a good candidate. The younger and more active southern Red Sea is more promising. Ron W. Girdler and Peter Styles, who were then working at the University of Newcastle upon Tyne, have attempted to identify magnetic anomalies in the Red Sea so that the sea-floor spreading there can be related to the spreading in the Gulf of Aden. The effort has been complicated by the intricate faulted structure of the floor of the Red Sea and by the slow rate at which it is opening. So far Girdler and Styles have been able to recognize two probable phases of spreading, one between 30 and 15 million years ago and then a recent one beginning five million years ago that is continuing today. Magnetic mapping by Hans A. Roeser of the Federal Institute of Geosciences and Natural Resources in Hanover reveals at the Red Sea a V-shaped zone of oceanic anomalies quite like the one in the Gulf of Aden. The magnetic quiet zone at the Red Sea appears to correspond to the older phase of spreading.

The tip of the putative rift that is opening the Red Sea is now between 15 and 16 degrees north latitude. Its distance from the tip of the Asal-Aden rift is 300 kilometers. Presumably that is the length of the Afar locked zone remaining to be broken apart. A comparison of Afar with the model described above suggests that the locked zone was originally 750 kilometers long and 150 kilometers wide. Each of the propagating rifts has therefore torn through 200 kilometers of the zone. An extrapolation of the current rates of propagation suggests that the locked zone will break up completely in less than 10 million years. The zone will by then have undergone a mean deformation that will have stretched it by 80 percent of its original width. The last part of the zone to rupture will have stretched twice that amount.

How do the findings at Afar bear on the suggestion by Wilson, Burke and Dewey that rifting is related to the doming and rupturing of the continental crust above a hot spot in the mantle? Afar has repeatedly been considered one of the clearest cases of a triple junction caused by a hot spot. The junction is that of the Red Sea rift, the Gulf of Aden ridge and the Ethiopian rift (the northernmost part of the East African rift). Our evidence, however, indicates that the rifts at Afar are propagating toward, not away from, an eventual junction. One of us (Courtillot) has suggested that a first phase of tectonic activity at Afar may have generated a knee-shaped rift that opened the Red Sea and left the East African Rift Valley. Preexisting zones of weakness and heterogeneity in the composition of the crust might be responsible for the "knee." A second phase of activity, still in progress today, may then have opened the Gulf of Aden and produced the Red Sea's second episode of spreading. Again a knee-shaped rift appeared, but since the Afar locked zone is not yet cut through, the rifting is not complete. The Afar junction would thus be the superposition of two knee-shaped rifts.

Rifts that are propagating through a continent and are breaking it apart will doubtless be found in places other than Afar. The Gulf of California is a likely candidate, even though its relatively small scale and its great disruption by transform faults make the magnetic data difficult to interpret. The process of rift propagation is well worth understanding. For one thing, the rifting of continental margins and the subsequent deposition of sediments above them govern the evolution of oil-bearing formations. In addition the study of rifting may yield insight into the mechanical properties of continental crust. The pattern of rifting that separated South America and Africa, for example, suggests the locked zones typically have a length of 400 kilometers, a width of 150 kilometers and are spaced at a mean interval of 700 kilometers. The pattern further suggests that sea-floor spreading begins (that is, oceanic crust begins to rise into place at a rift) when about 65 percent of the length of the future plate boundary has lost its strength to resist the breakup.

In attempting to understand how a plate breaks up over a length of thousands of kilometers, we have dealt with temporal scales of no more than a million years and spatial scales of hundreds of kilometers. On such scales we find the continental crust deforming. Among the tools that facilitate the study of how the deformation proceeds two are crucial: the detailed examination of magnetic anomalies and the detailed examination of the edges of the continents. To the extent that the work goes beyond the assumption that plates are rigid it should provide a better view of how a plate evolves.

Oceanic Fracture Zones

They complicate the pattern discerned in the theory of plate tectonics by dissecting the edges of the plates that compose the ocean floor. Some of them are the width of an ocean basin.

• • •

Enrico Bonatti and Kathleen Crane
May, 1984

The realization that the ruggedest topography on the surface of the earth is on the floor of the oceans is little more than two decades old. First the midocean-ridge system was discovered. It is a submarine chain of mountains that traverses the major ocean basins for a length of some 60,000 kilometers, thus constituting by far the longest mountain range on the earth. Then a compilation of topographic data for the major ocean basins revealed a complication in their geometry: the axis of the midocean ridge is offset laterally in many places by distances ranging from a few kilometers to several hundred. The offsets are particularly common along the Mid-Atlantic Ridge: if one traces the crest of the ridge, one encounters offsets at intervals of 50 to 100 kilometers. Most of them are short: one must travel less than 30 kilometers to find where the ridge continues. Some are notably longer. Characteristically each of the longer offsets consists of a deep trough joining the tips of two segments of the ridge. The trough is bounded by elevations running more or less parallel to the trough. They are called transverse or transform ridges. Remarkably, the trough and its flanking ridges can often be traced well beyond the ridge-axis segments they join. The

trough and the transverse ridges thus form extensive ocean-floor disruptions now known as an oceanic fracture zone.

Consider the equatorial Atlantic. Here a set of closely spaced fracture zones dissects the Mid-Atlantic Ridge. The largest of them, the Romanche Fracture Zone, offsets the axis of the ridge by almost 1,000 kilometers (see Figure 6.1). The deepest parts of the floor of the Romanche trench are more than seven kilometers below sea level; the highest parts of the ridges flanking the trench are less than one kilometer below sea level. Hence the vertical relief is more than six kilometers. The Grand Canyon is scarcely a fourth that deep. The Romanche Fracture Zone is flanked by several similar zones that are almost equally impressive, producing a sequence of troughs and transverse ridges several hundred kilometers wide from north to south. The resulting terrain, the Equatorial Megashear Zone, is hardly matched in size and ruggedness anywhere else on the planet.

What geologic processes created the fracture zones of the equatorial Atlantic and similar fracture zones discovered elsewhere in the earth's ocean basins? The answer requires that the events produc-

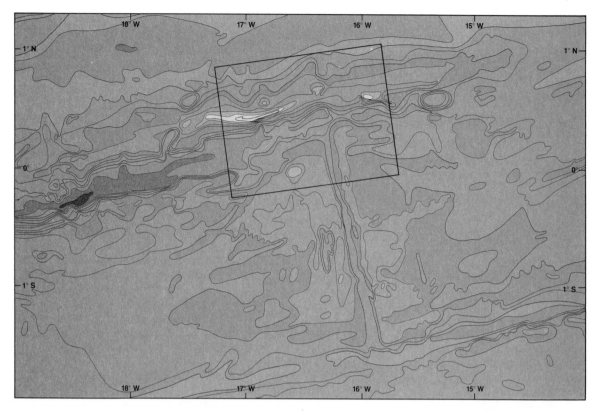

Figure 6.1 ROMANCHE FRACTURE ZONE is the most pronounced of the fracture zones composing the Equatorial Megashear; its depth varies from less than one kilometer at the top of its ridges to more than 7.5 kilometers at the deepest parts of its trough. Here detailed topographic relief for part of the zone is mapped. Contour lines are at 500-meter intervals from 1,000 to 5,000 meters, with additional contours at 6,000 and 7,000 meters. The characteristic pattern is clear: a prominent trough is flanked by prominent ridges. A segment of the Mid-Atlantic Ridge meets the zone from the south. The part of the fracture zone enclosed by the rectangle appears in Figure 6.8.

ing an ocean basin be examined. The fracture zones then emerge as places of massive geologic activity. The zones arise as part of the process by which an ocean basin opens. Then at later times they are involved in massive readjustments.

Perhaps the first clue to the nature of fracture zones was the finding that the crest of the mid-ocean ridge is a locus of intense seismic and volcanic activity and of a great flow of heat from the interior of the earth. This finding and others were explained by the theory of plate tectonics. One of the tenets of the theory is that new oceanic crust is created along the crest of the midocean ridges and is added to the great plates that form the floor of the ocean basins. The plates carry crust away from the crests at speeds ranging from one centimeter per year to 20

centimeters per year. The crests are thus the axes of sea-floor spreading. In some cases, including that of the Atlantic, the motion of the oceanic plates pushes continents apart, providing a mechanism for continental drift.

The geologic significance of the lateral offsets in a crest, however, was unclear until 1965, when J. Tuzo Wilson of the University of Toronto introduced the concept of the transform fault into the hypothesis of sea-floor spreading. Within that framework one oceanic plate carries crust away from one side of the crest of a midocean ridge and another plate carries crust away from the other side (see Figure 6.2). Therefore along an offset of two ridge-axis segments, blocks of crust would be sliding past each other as they were carried in opposite directions. The offset is a transform fault: a zone of

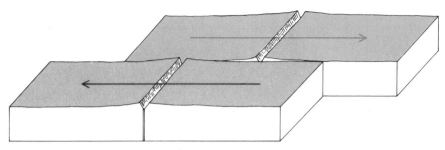

Figure 6.2 FIRST HYPOTHESIS about oceanic fracture zones was that they are much like a continental transcurrent fault: they are lines along which two blocks of crust slide past each other. According to this hypothesis, segments of a midocean ridge are carried along with the blocks.

what geologists call strike-slip motion (see Figure 6.3).

In Wilson's model the strike-slip motion occurs at the length of the transform and not beyond it, along the line representing the projection of the transform fault across an ocean basin. That prediction is borne out by the observation that earthquake epicenters are common within the transform but uncommon along its projection. In addition the prediction is supported by an analysis of earthquake seismic waves done by Lynn R. Sykes of the Lamont-Doherty Geological Observatory of Columbia University. The waves reveal the direction in which crustal blocks abruptly slip, generating earthquakes in the transform zone. Presumably the direction of slippage is also the direction of the slower tectonic motion. Sykes's analysis indicates that the motion is indeed strike-slip and that it follows the directions Wilson predicted.

Clearly Wilson's model of oceanic transform faults is essentially correct. Nevertheless, recent research suggests it is an oversimplification. For one thing, the trough and the transverse ridges that are the signature of large-offset transform faults can be traced outside the offset zone, along the projection of the offset. Indeed, the signature of some equatorial-Atlantic transforms can be traced from one side of the Atlantic to the other, that is, from the coast of Africa to the coast of South America (see Figure 6.4). Near the margin of each continent it is buried under sediment, but seismic surveys reveal it is there. The concept that transform faults are places where crustal blocks slide past each other more or less passively is yielding, therefore, to the concept that they are places of complex geologic activity of prime importance in determining the structure and evolution of ocean basins.

What determines the presence of a deep trough along a transform fault? Several factors are probably at work. Fundamentally, new oceanic crust forms along the axis of a midocean ridge be-

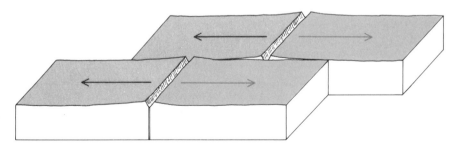

Figure 6.3 PLATE-TECTONIC THEORY proposed that new oceanic crust forms along each segment of a midocean ridge. Thus the ridge is stationary, and a plate diverges from each side of it. Two plates slide past each other only along the offset between the two ridge-axis segments.

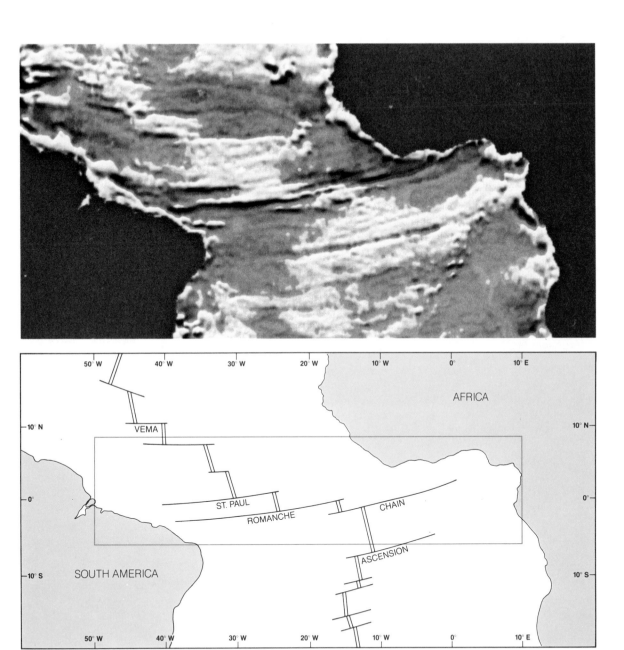

Figure 6.4 SEA-LEVEL MEASUREMENTS made by radar from a satellite yielded this image of a series of fracture zones dissecting the Mid-Atlantic Ridge. White patches mark areas such as the ridge itself, where the sea floor is relatively shallow. There the ocean's level rises. Dark patches mark increasing depths, notable among which are deep troughs found to be associated with oceanic fracture zones. There the ocean's level sinks. Many of the zones traverse the width of the equatorial Atlantic. The zones are identified in the accompanying map.

cause an upwelling of hot basaltic magma, produced in the mantle of the earth some 30 to 60 kilometers below the sea floor, cools and solidifies at the axis of the ridge as it reaches the ocean floor. The level attained by the magma corresponds to the depth of the axis (that is, its distance below sea level) and is governed by the quantity of the magma, by its temperature and viscosity and by the drag due to the interaction of the magma with the relatively cold walls of the conduit where the upwelling takes place. The crust moves away from the axis, and as it moves it continues slowly to cool and contract. As a result the depth of the ocean floor increases both with the floor's age and with its distance from the axis at which it formed. The relation is expressed by a law that is gratifyingly simple and has been well verified in the three major oceans, the Atlantic, the Pacific and the Indian Ocean: the depth of the ocean increases with the square root of the age of the crust (see Figures 2.4 and 2.6).

At a transform fault this simple law is broken: the trough is in essence an anomalous oceanic deep. At a transform fault, however, a hot young ridge-axis segment meets older, colder lithosphere. (The lithosphere includes the crust and the relatively rigid upper part of the mantle. Together they make up the thickness of a tectonic plate.) Thus the hot material upwelling along the axis near a transform fault must interact with a cold surface it does not encounter elsewhere. It must then "freeze" at a deeper level, producing a topographic low at the intersection of the axis and the transform. The spreading of the plates extends the topographic low into a trough that can traverse the entire width of an ocean basin. The trough may even widen during the initial stage of sea-floor spreading, because the cooling of a plate presumably causes it to contract not only vertically but also horizontally.

The transverse ridges flanking the trough constitute a further infringement on the law linking the depth of the ocean to the age of the oceanic crust. They are anomalous highs. How, then, do they originate?

The rocks forming the walls of transform troughs and the slopes of transverse ridges have been sampled extensively in the past decade (see Figure 6.5). Most of the samples have been recovered by scratching the sea floor with a dredge maneuvered from a ship; some were recovered by the submersible research vessel *Alvin* in the course of a study of the Oceanographer Transform Fault (done by us along with Paul J. Fox of the University of Rhode Island and workers from the State University of New York at Albany and from the Woods Hole Oceanographic Institution). The Oceanographer transform is at 35 degrees north latitude, near the Azores in the North Atlantic.

A mong the rocks that were sampled two types turned out to be abundant. The first was peridotites (see Figure 6.6). Mineralogically peridotites are rocks consisting chiefly of olivine, pyroxenes and spinels. Olivine and pyroxenes in turn are silicates of elements such as magnesium and iron; spinels are oxides of magnesium, aluminum and chromium. Thus the magnesium content of a peridotite (expressed as a net content of magnesium oxide) is high: it can reach 40 percent. The magnesium content of a basalt (the characteristic rock of the oceanic crust) is no more than a fourth as great. Physically peridotite is a dark greenish rock with a high density, typically 3.2 grams per cubic centimeter. That density is consonant with the hypothesis that peridotites are the chief constituent of the upper mantle. The density of the upper mantle has in fact been inferred from the velocity at which seismic waves travel through it. The result suggests that the density of the upper mantle under the oceans is from 3.1 grams per c.c. to 3.3.

The second abundant rock was gabbros. Mineralogically and chemically they are much like basalt: they consist chiefly of silicates of calcium and aluminum, which form minerals such as feldspar. Physically too they resemble basalt: they are grayish and have a density of about 2.8 grams per c.c. The principal difference between a gabbro and a basalt is that the crystals in gabbro are larger. It is therefore inferred that gabbros and basalt both arise from the partial melting of peridotites in the mantle. The parts of the melt that reach the sea floor cool rapidly when they erupt. The result is basalt. The other parts cool slowly at greater depths. The result in that case is baggro. Specifically gabbro solidifies under the axis of the midocean ridges, in magma chambers in the lower part of the oceanic crust.

The crucial point is that both types of rock abundant at transform faults—peridotite and gabbro—are normally deep-seated. Peridotite is in the upper mantle, gabbro is in the lower crust. The presence of these rocks suggests, therefore, that transverse ridges are slivers of deep crust and upper mantle that have somehow been uplifted. In this regard the study of sections of transverse ridges made up en-

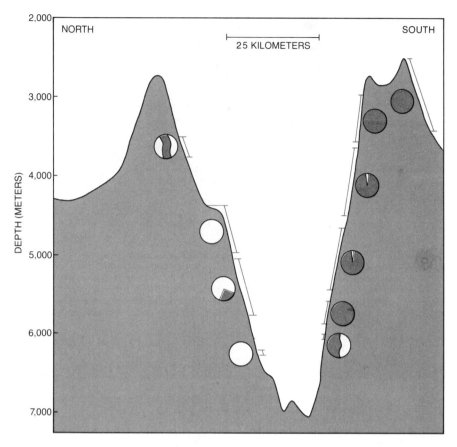

Figure 6.5 PROFILE ACROSS ROMANCHE shows locations at which rocks were dredged by the authors. A small pie chart at each location shows the identity of the rocks. Some of the rocks (*white*) were unexceptional; they included basalt, the rock of which the oceanic crust consists. On the other hand, peridotite (*color*) proved to form 4.5 kilometers of the height of the southern wall of the Romanche trough. Peridotite is not normally part of the oceanic crust. Note that the vertical dimension has been exaggerated about twenty-five times that of the horizontal dimension.

tirely of mantle-derived peridotite has been particularly revealing. The minerals in peridotite are stable only at the great temperatures and pressures of the mantle. Elsewhere they tend to be changed into other minerals. Moreover, some of the minerals in peridotite, particularly pyroxenes and spinels, vary slightly in chemical composition depending on the temperature and pressure—that is, the depth—at which they form. The mineral assemblages in the peridotites sampled at transform faults suggest they derive from depths in the earth exceeding 30 kilometers. (The boundary between the crust and the mantle is generally at a depth of from four to five kilometers.) These peridotites must somehow have risen that distance as essentially solid bodies.

In the course of such an ascent the peridotite will have reached levels in the crust, a few kilometers below the sea floor, where seawater is particularly likely to penetrate into the crust at midocean ridges and transform faults because the crust in such places is greatly fractured and therefore is permeable. Some of the minerals in peridotite (notably olivine and pyroxenes) can react with the seawater, so that they are gradually changed into other minerals, notably the series of hydrated magnesium silicates called serpentines (see Figure 6. 7). As a result

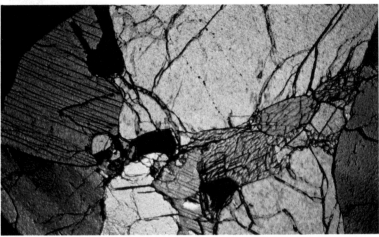

Figure 6.6 PERIDOTITE from the Oceanographer Fracture Zone in the North Atlantic was photographed from the submersible research vessel *Alvin* (*left*); its internal structure is suggested by that of a peridotite from a Red Sea fracture zone, photographed in thin section illuminated by polarized light (*right*). Peridotite consists of three minerals: pyroxene, olivine and spinel. A crystal of pyroxene (*brown in this illumination*) is at the upper left; crystals of olivine (*green and orange*) and spinel (*black*) fill the rest of the field. The field is about .5 centimeter across. Peridotite is thought to occupy the upper part of the mantle, the layer subjacent to the crust. Hence its presence in the elevated parts of fracture zones suggests that the elevations are slivers of crust and upper mantle lifted through vertical distances as great as 30 kilometers.

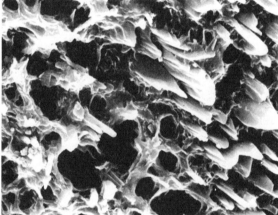

Figure 6.7 SCANNING ELECTRON MICROGRAPHS of peridotites from oceanic fracture zones show how they can be altered by contact with seawater. The peridotite at the left is from the Red Sea. It is unaltered. The surface of a crystal of pyroxene is stepped by a series of cleavages. The peridotite at the right is from Romanche. It is greatly altered: contact with seawater percolating downward through cracks in the oceanic crust has hydrated the pyroxene in it, forming a network of crystals of the mineral serpentine. The hydration and the porosity of the network have markedly reduced the density of the peridotite, perhaps facilitating its ascent from the upper mantle.

the density of the peridotite decreases dramatically, from about 3.2 to 2.6 or even less. The decrease must tend to facilitate the ascent of the rock through the denser materials of the crust, including gabbro and basalt.

This is not to say the uplift of deep-seated rock at oceanic fracture zones has been inferred only from subtle analyses of rocks recovered from great depths in the ocean by dredging or with submersible research vessels. In at least one case the summits of a body derived from the mantle reach above sea level: the St. Peter-Paul islets rise from the deep ocean floor in the middle of the Atlantic just north of the Equator. In 1831 Charles Darwin, aboard the *Beagle*, landed briefly on their desolate, battered terrain. He saw that the islets were geologically unusual: they appeared not to be volcanic. It is now known that the islets consist of peridotites. They are a fragment of upper mantle that has been uplifted close to the intersection of the St. Paul transform with the axis of the Mid-Atlantic Ridge.

Moreover, it is now recognized that parts of other transverse ridges in the equatorial Atlantic have emerged as islands in recent times, geologically speaking. At the Vema transform, 11 degrees north of the Equator, the axis of the Mid-Atlantic Ridge is offset by about 300 kilometers. Throughout the length of the offset an uplifted silver of oceanic crust lies on the south side of the transform trough. The summit of the uplifted silver is only 600 meters below sea level. A detailed sampling of the sliver shows it is capped by reef limestones similar to those ringing islands such as the Bahamas.

Several lines of evidence have helped us to deduce the environment in which the limestones formed. For one thing the limestones include shallow-water structures such as oolites: calcareous deposits formed when cold water flows onto warm, shallow banks. They include the fossil remains of organisms such as corals, calcareous algae and mollusks, which live in shallow water or in the zone between high tide and low tide. In addition the limestones show alterations typical of erosion on land. Finally, the proportions of the isotopes of oxygen are revealing. If a marine organism forms a shell of calcium carbonate, the oxygen atoms in the $CaCO_3$ will have a ratio of the isotopes oxygen 16 and oxygen 18 related to that of the oxygen dissolved in seawater. If at a later time the shell is exposed to rainwater, in which the ratio of the isotopes is slightly different, the ratio will change.

The evidence confirms that the limestones formed close to sea level as part of coral reefs and lagoons. Moreover, the evidence shows that the limestones were weathered and compositionally altered under the sky. This means that the shallowest parts of the Vema Transverse Ridge emerged as an island or islands at some time in the past. The evidence tells when. The identification of the fossil shells of foraminifera, a microscopic marine organism, suggests that the Vema Transverse Ridge was above sea level either continuously or intermittently from mid-Miocene to mid-Pliocene times, that is, from about 10 million to three million years ago. It has since sunk into the sea at an average rate of .3 millimeter per year.

A similar situation is found at the eastern intersection of the Romanche transform with the Mid-Atlantic Ridge (see Figure 6.8). Here the Romanche Transverse Ridge shallows to little more than 1,000 meters below sea level for a distance of more than 100 kilometers. Again we were able to demonstrate that the summit of the ridge is capped by a fossil reef and that the summit emerged some five million years ago as an island or series of islands. They have subsided since then at the average rate of .2 millimeter per year.

At the risk of oversimplifying, let us consider quite schematically what would happen at an oceanic transform fault if the direction of spreading of the oceanic plates that slide past each other along the length of the transform were to change even slightly from the orientation in which the transform makes a right angle with each segment of a ridge axis (see Figure 6.9). Suppose the angles increased. The plates would no longer slide past each other. Instead they would diverge. The transform zone would be subjected to extension; thus the zone would tend to open. For example, the zone might widen, and into the extended, weakened part of the oceanic crust some basaltic magma could rise. This possible formation has therefore been termed a "leaky" transform. Conversely, the angles could decrease. The plates would then converge on the transform zone. The crust in the zone would be deformed by lateral compression, and the resulting vertical movements could conceivably bring on the tectonic uplift of slivers of crust.

Is there any evidence that the direction of spreading of oceanic plates has changed over time? More than a decade ago Henry W. Menard, Jr., and Tanya M. Atwater of the Scripps Institution of Oceanogra-

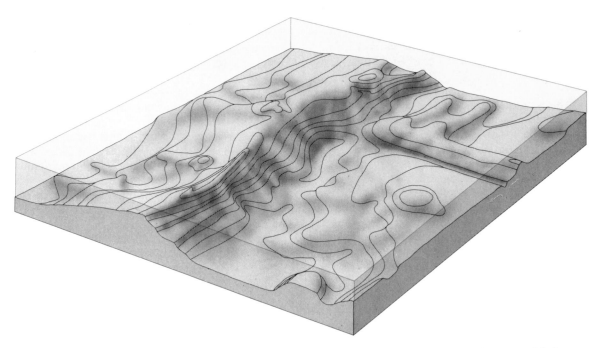

Figure 6.8 OBLIQUE VIEW of part of the Romanche Fracture Zone emphasizes the roughness of its topography by exaggerating the vertical scale. Contour lines are at 500-meter intervals from 1,000 to 5,000 meters. In plate-tectonic theory the elevation of the sea floor decreases with the age of the crust. The trough and ridges that characterize an oceanic fracture zone are thus anomalies in the age-depth relation.

phy pointed out that magnetic anomalies in several parts of the earth's ocean basins do show changes in orientation. Magnetic anomalies are perhaps the most important evidence favoring the theory of plate tectonics. They arise because oceanic crust forms from cooling lava, and the magnetic field in the lava becomes aligned with the field of the earth at the time. The earth's field has reversed its polarity many times throughout geologic history; hence the field in the crust that formed at a midocean-ridge axis and then was carried off by sea-floor spreading should bear the record of a series of reversals. It does. Moreover, the pattern of reversals on one side of a midocean-ridge axis is the mirror image of the pattern on the other side of the axis. What Menard and Atwater saw is that the orientations of the magnetic "stripes" between reversals do not always parallel the orientations of the midocean ridges. The implication is that the direction of sea-floor spreading can change.

A second piece of evidence suggesting that the direction of sea-floor spreading can change is provided by the oceanic fracture zones themselves. The troughs and transverse ridges produced at transform faults and carried across the ocean basins by sea-floor spreading display sharp breaks in orientation. The breaks are clearly seen in the maps of the floor of the Atlantic produced with radar altimetry by William F. Haxby, a colleague of ours at Lamont-Doherty (see Figures 2.4, 2.6, and 3.1). Presumably each break was caused by the stresses arising when a plate changed its direction of motion.

Qualitatively, then, we could assert that the vertical tectonic movements giving rise to the anomalous shallows at transform zones are caused primarily by compressional stresses resulting from the reorientation of the transform fault in response to changes in a plate's direction of motion. The data that would make this assertion quantitative are not yet all available. One would need measurements of strain in the crust and also an extensive recovery of rock samples.

The closest we can come at present to making a

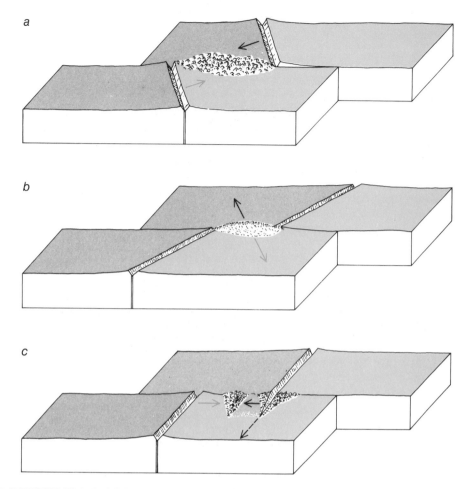

Figure 6.9 STRESSES IN A FRACTURE ZONE may arise when plates change their direction of spreading. If the direction changes so that plates converge at the fracture zone (*a*), the crust may be compressed, causing the sea floor to rise. If the direction changes so that plates diverge from the zone (*b*), the crust may be stretched and weakened, forming a "pull-apart basin." The propagation of the tip of one of the ridge-axis segments (*c*) may also cause plates to be compressed.

quantitative analysis is at the Vema Fracture Zone (see Figure 6.10). There in the early 1970's T. H. Van Andel of the Scripps Institution recognized the presence of a set of inactive transform valleys to the south of the transform that is known to be active today. Each inactive valley makes an angle of from 10 to 15 degrees with respect to the active transform. In addition each inactive valley can be traced only in oceanic crust more than 10 million years old.

These observations can be interpreted according to the following model. Before 10 million years ago a transform fault was active south of the present-day Vema transform (see Figure 6.11). The tip of one of the Mid-Atlantic Ridge axis segments joined by the fault then propagated northward. Longitudinal propagation of that type has been documented in the East Pacific, in the Gulf of Aden, in the Red Sea and elsewhere by investigators including R. N. Hey of the Scripps Institution and Vincent Courtillot of the University of Paris. Next the Vema transform relocated so that the fault and the axis segments it joins assumed the geometry now ob-

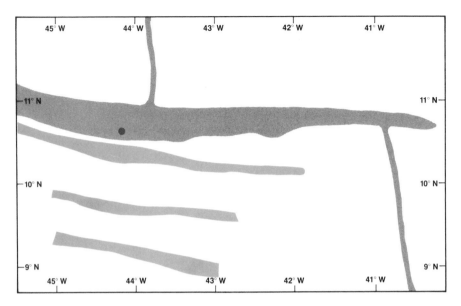

Figure 6.10 MAP OF THE VEMA FRACTURE ZONE in the Equatorial Megashear shows the presence of numerous valleys (presumably inactive transform faults) south of the transform that is active today. The map displays the work of T. H. Van Andel of the Scripps Institution of Oceanogra- phy. In part of the Vema zone (*colored dot*) the rock proved to be limestone some 55 million years old. If the limestone had always been south of the Vema transform and had been carried to its present-day position by westward sea- floor spreading, it would be only half that age.

served. It is a geometry in which the fault meets the axis segments at right angles and the stresses in the region are relieved.

The model accounts for a curious finding we made at Vema: the uplifted crustal block on the south side of the current transform fault is anoma- lously old. In particular some of the limestone re- covered from the block proves to be more than 55 million years old. (It is limestone representing not a fossil reef but the solidification of calcareous sedi- ment on the ocean floor.) Yet the rate of sea-floor spreading at Vema is little more than one centimeter per year. If the basaltic crust underlying the lime- stone had formed at the ridge-axis segment south of the current Vema transform and then had been moved, with the limestone on top of it, to its present location, the limestone should be no older than 30 million years.

We suggest that the crustal block capped by the limestone has undergone a process we call oscilla- tory spreading. Before the Vema transform relo- cated, the limestone was north of the fault and was moving eastward as part of the African plate. After the relocation (and within the present geometry) it is south of the fault and is moving westward as part of the American plate. Because of the reversal in its direction of travel, it has been moving for more than 55 million years but has yet to leave the area. Pre- sumably the fault's transition from one configura- tion to the other was accompanied by the strong compressional stresses that lifted the crustal block the limestone is on.

The observations and inferences at the Vema Fracture Zone demonstrate that the long trans- form faults dissecting a midocean ridge cannot be seen as quiescent features of the ocean floor. They are notably active features: they are subject to mi- gration, to change in orientation and to deforma- tion. One can see why. A length of several hundred kilometers means that a transform fault traverses relatively cold oceanic crust as it passes from one ridge-axis segment to another. The crust there be- haves rather rigidly. If the offset were shorter, the crust it traverses would be hotter and more plastic. Readjustments would be easier. Perhaps it is little cause for wonder that topographic roughness and seismicity both increase with the length of the offset between two ridge-axis segments.

The transforms dissecting the Mid-Atlantic Ridge

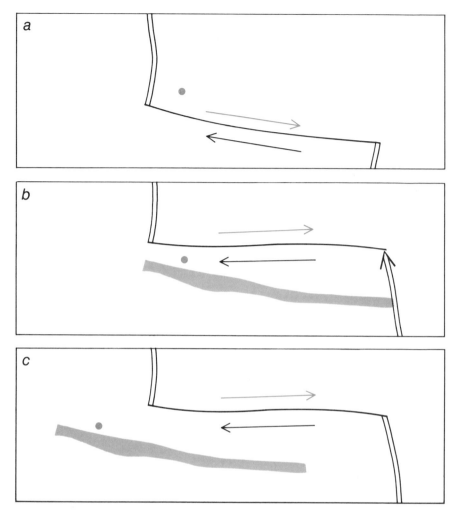

Figure 6.11 OSCILLATORY SPREADING is proposed as the reason for the anomalous age of the limestone sampled at Vema. At first the limestone (*colored dot*) was north of the Vema transform fault and so was moving eastward as part of the African plate (*a*). Then, however, the position of the transform shifted, so that the limestone is south of it (*b*). Today the limestone is moving westward as part of the American plate (*c*). A fossil valley marks the transform's former position.

are generally much rougher than the ones dissecting the East Pacific Rise: the great ridge axis of the Pacific. Again one can see why. Sea-floor spreading is up to 10 times faster at the East Pacific Rise than it is in the Atlantic. Thus for equal lengths of offset the crust affected by a transform will be younger, hotter and less rigid in the Pacific than it is in the Atlantic.

Ultimately, of course, the instability of ridge-axis transforms must occur in response to the distribu-tion of stresses within the oceanic plates, and the stresses must occur in response to the large-scale movements of the plates. The instability must there-fore be extreme when a continent breaks up and an ocean basin opens. After all, continental lithosphere is thicker, colder and more rigid than oceanic litho-sphere. The opening of the Atlantic is a case in point.

An analysis of the pattern of magnetic anomalies on the floor of the present-day Atlantic reveals that the Atlantic ocean basin broke open in discrete in-crements. Moreover, an analysis of the basin's geol-ogy suggests that in both the North and the South Atlantic the initial lines of opening were determined by the position of preexisting fracture systems tra-versing the continent that was soon to break apart. Presumably these zones were the wounds of tec-tonic events (collisions between plates, say) that oc-curred before the Atlantic developed. From the south to the north the major fracture systems were the Falkland Shear Zone and the Equatorial Mega-shear Zone, dominated by the Romanche Fracture Zone; then, in the North Atlantic, the Atlas Shear Zone and the Charlie Gibbs Shear Zone; and finally, in the Norwegian Greenland Sea, the Jan Mayen and the De Geer shear zones. The evidence suggest-ing the presence of the Equatorial Megashear Zone is particularly good. The remnants of the zone can be recognized in the Amazon trough on the Ameri-can side and in the Benue trough on the African side.

The South Atlantic opened, then, from the south to the north, like a zipper, at a rate that varied from 10 to 20 centimeters per year, according to work by Philip D. Rabinowitz and John L. LaBreque of Lamont-Doherty (see Figure 6.12). In essence the tip of a midocean-ridge axis moved northward, giv-

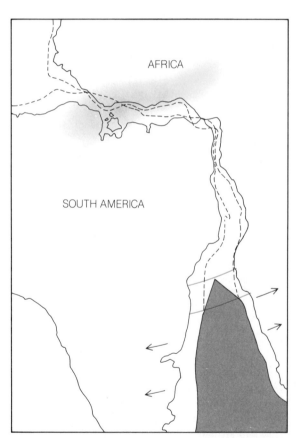
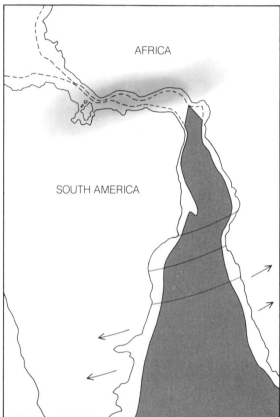

Figure 6.12 OPENING OF THE ATLANTIC was guided by preexisting fracture zones in the continental crust. By 105 million years ago (*left*) the northern tip of a propagating rift had reached the first shear zone in the South Atlantic. Then, five million years later (*right*), the tip of the rift reached the Equatorial Megashear (*color*). At each succes-sive encounter the northward advance of the rift was im-peded until enough newly minted oceanic crust had spread from the rift to allow the fracture zone and the rift to reorient themselves in a direction at right angles to the spreading. (See Figure 6.13 for hypothetical details of the reorientation.) The ancient positions of Africa and South America are inferred by repositioning their present-day outlines (*solid lines*) and their continental shelf, defined by the oceanic 1,000-meter contour (*broken lines*).

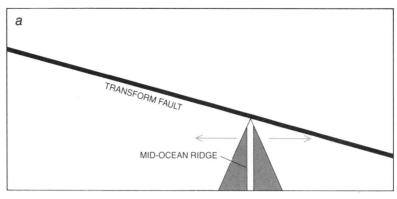

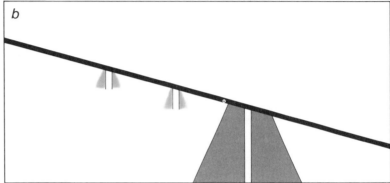

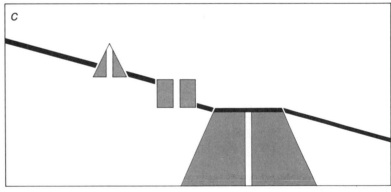

Figure 6.13 HYPOTHETICAL SE-QUENCE illustrates how a propa-gating rift might impinge on a preexisting shear zone. The illus-tration assumes that a propagating rift such as the one that opened the South Atlantic would tend to meet a preexisting shear at an oblique angle (*a*). Hence at one side of the rift the newly formed oceanic crust would be compressed against the shear; at the other side the crust would be stretched and would weaken. The weakened crust would admit basaltic magma; thus new oceanic spreading centers would arise (*b*). Ultimately they would produce enough oceanic crust for the fracture zone to relocate (*c, d*). In the illustration one of the new spreading centers resumes the propagation, so that the ridge axis comes to have offsets in it.

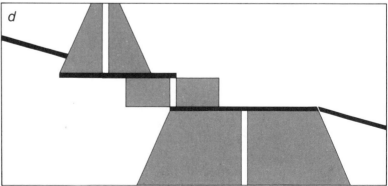

ing rise to oceanic crust that spread to each side. The motion began somewhere along the Falkland Shear Zone about 130 million years ago.

By 100 or 90 million years ago the tip of the axis had reached the Equatorial Megashear Zone (see Figure 6.13). The encounter must have been dramatic. For one thing, the Equatorial Megashear may well have represented what Courtillot calls a locked zone, that is, a part of a plate greatly resistant to being opened by a propagating rift (see Figure 5.9). Then too, it is likely that the propagating axis tip impinged on the Megashear at an angle other than 90 degrees. The Megashear and the rift arose in different tectonic episodes, and so there is no reason to suppose they would be at right angles to each other. The oblique configuration would have been highly unstable. At one side of the ridge axis newly formed oceanic crust would have rammed into the Megashear, causing intense compression and probable uplift. At the other side of the axis the opposite would have happened: the stress would have been extensional. Perhaps the upwelling of basaltic magma into the weakened crust created a series of local oceanic spreading centers there.

Meanwhile the tip of a North Atlantic rift was apparently moving southward toward the Equator. The lateral distance from it to the southern rift would have been so great—more than 2,000 kilometers—that in order for the Atlantic to open at the Equator, a series of long-offset transforms would have had to develop. For at least 20 million years these transforms would have lain in continental lithosphere. Moreover, they probably would not have been at right angles to the rifts. Intense stresses must have arisen. The stresses would have continued until the quantity of newly minted oceanic crust enabled the rift segments and the fracture zone connecting them to readjust to a more nearly right-angled configuration, one that matched the direction of relative plate motion. On the basis of work by Dennis E. Hayes of Lamont-Doherty we surmise the new ocean basin would have had to be

at least 1.5 times the width of the offset traversed by the fracture zone in order for the entire Equatorial Megashear to have made the readjustment. Only then could the zone have migrated and rotated with little or no impedance from continental lithosphere.

Surely the opening of the equatorial Atlantic was greatly influenced by the Equatorial Megashear Zone. One result is that the Mid-Atlantic Ridge axis shows the impressive series of long offsets observed at the Equator today. A second result is the general shape of the Atlantic. The rifts that opened the ocean tended to separate Europe and Africa from the Americas along a roughly north-south line. At the Equator, however, the Megashear Zone made the line east-west. Traces of the initial separation can be seen in the steplike east-west offsets of the equatorial Atlantic coastlines of both Africa and South America.

Throughout the further evolution of the Atlantic the equatorial transforms have been active, so that each change in the direction of sea-floor spreading has had dramatic consequences. In each case the readjustment of the geometry of the equatorial transform faults will have led to episodes of extension and compression, giving rise to volcanism and to the vertical motion of crustal slivers. On one resulting uplift along the St. Paul transform the birds observed by Darwin build their nests on the mantle of the earth. Elsewhere fossil coral reefs have sunk to the depths of the sea. It is likely that in the young and narrow equatorial Atlantic crustal blocks rose and fell quite often. Thus the geology of oceanic fracture zones probably influenced events as diverse as migrations of life between the continents and exchanges of water between the northern and southern ocean. The poetic intuition of William Carlos Williams was keen when he wrote:

Oh most powerful connective, a bead
to lie between continents through
which a string passes. . . .

III

MAKING MOUNTAINS

Introduction

Despite their complexity, modern oceanic plate boundaries are still relatively simple when compared with the orogenic regions of continents, as described in Chapter 1, "The Continental Crust." Active continental orogenic regions are mountainous areas where oceanic plates descend beneath continental margins, such as the Andes; where continents are colliding with each other, such as in the Himalayas, or with oceanic island arcs, as in New Guinea (see Figures 1.2 and 1.8). Also, as these continental regions contain most of the surviving information about the earth's tectonic activity for the first 95 percent of its history, their understanding is an essential part of our attempt to understand the overall tectonic development of the earth.

Chapter 7, "Terranes," discusses their role in tectonics. Terranes are fault-bounded regions of crust with a distinctive and separate history that have been incorporated into orogenic belts. The present oceanic crust includes not only average oceanic crust, but many regions of anomalously thick crust as well. These latter regions include traces of hot spots, such as the Hawaiian chain and its older continuation, the Emperor seamount chain; volcanic plateaus, such as the Faeroes and Vøring plateaus in the northern Atlantic; ocean island arcs formed as one oceanic plate descends beneath another, such as the Marianas, the Tonga Kermadec chain or the Lesser Antilles, and even small bits of continent left behind as major continents rift apart and separate, such as the Seychelles and Madagascar.

As ocean plates that contain such anomalous areas of crust descend beneath continents, and as continents themselves ultimately converge and collide, the areas of anomalous crust are swept into the edge of the continent above the subduction zone. We can see these regions along such active or recently active continental margins as terranes attached to the continent.

The recognition of the importance of these terranes along active continental edges is a relatively new development in the understanding of the growth of continents and development of orogenic belts. It now appears, however, that such terranes are important features of all orogenic belts as old as two billion years.

Chapter 8, "Ophiolites," discusses these distinctive sequences of rocks found in the world's orogenic belts that seem to represent fragments of ocean crust and mantle formed at spreading centers and subsequently incorporated onto continental edges. The existence of these sequences of rocks has been known since the start of the 20th century, but their true significance was not recognized until the late 1960's.

Ophiolites are important for three reasons. First, they demonstrate that the orogenic belt in which they reside includes the remnant of ocean crust formed at a spreading center. Ophiolites within an orogenic belt within a major continent, such as the Urals, which separate more ancient rocks in Europe from those of Asia, are themselves evidence that an ocean basin once existed between two formerly separate and now joined continental masses.

Second, many well-preserved ophiolites provide insight into the nature of the oceanic crust formed at spreading centers. In Chapter 2, "The Oceanic Crust," we discussed the difficulty in obtaining a view of the oceanic crust presently lying beneath the world's oceans. Ophiolites are well-exposed on land and easy to see. In well-preserved examples it is possible to see the rocks and structures that form in the crust and upper mantle as plates spread.

Third, many major ophiolite complexes possibly were emplaced on continents by collisions of a continental margin with a subduction zone, such as the collision in New Guinea (see Figure 1.8). Ophiolites thus emplaced represent the remnant edges of the overriding plates and the faults beneath the ophiolite are remnants of subduction zones.

Since 1982 a great deal of progress has been made in the understanding of ophiolites. Much work on the Troodos complex in Cyprus and the Samail complex in Oman has refined our ideas on how the oceanic crust forms at spreading centers. Many more complexes have been discovered recently in such diverse regions as Norway, Nevada, northern

Labrador and the Karelian region of Finland. The recognition of these complexes enriches our understanding of the development of the orogenic belts that contain them.

Chapter 9, "The Structure of Mountain Ranges," uses as special examples the Andes and the Himalayas. As oceanic plates descend beneath continental margins, as in the Andes, or as two continents collide, as in the Alpine-Himalayan mountain system, major thrust faults form. This thrusting causes the crust of the continents to thicken, which in turn causes the surface of the earth to rise to higher elevations. This rise in the land surface coupled with the resultant increase in erosion by wind, water and ice, produces mountains as we know them.

Chapter 10, "The Southern Appalachians and the Growth of Continents" discusses the first of two case studies of orogenic belts. The southern Appalachians, which 200 to 300 million years ago formed a high mountain chain similar to the Himalayas today, formed by convergence and collision of the ancient North American continental edge first with a series of oceanic terranes, similar to those described in Chapters 7 and 8, and then with the margin of Africa. In the course of this tectonic history, structures formed that are reminiscent of those forming today in the Himalayas, as described in Chapter 9. Subsequent erosion of the resultant mountains has exposed deeper levels of the crust in the Appalachians than are yet apparent in the Himalayas.

The focus of the Appalachian discussion is the surprising result of a geophysical survey across the belt from Tennessee to Georgia, which revealed that all the rocks exposed at the surface are part of a thin sheet less than 15 kilometers thick underlain by a series of horizontal layers that are thought to be relatively undeformed sediments. Apparently the highly deformed sedimentary, metamorphic and igneous rocks at the surface were thrust westward over horizontal sedimentary rocks in the last stages of the tectonic movements that gave rise to the belt.

Since publication of the results in the southern Appalachians, similar work has found essentially identical structures in New England, in eastern Canada, in the predrift continuation of the Appalachian belt in Britain and Scandinavia, in the Alps and in the mountains of western Canada and Nevada. These surprising new results imply that similar large-scale thrusts of whole orogenic belts may

be widespread. If verified, this implication would considerably change our ideas of the scale of thrust tectonics involved in orogenic belt development. In addition this new result could have important bearing on the search for ever-scarcer hydrocarbon deposits.

Chapter 11, "The Growth of Western North America," presents a discussion of the tectonic development of that region in terms of the arrival and collision of terranes. Paleontologic and paleomagnetic investigations suggest that many of the terranes of western North America are far-traveled. Some may have originated from the Southern Hemisphere. Others represent fragments of oceanic islands or island arcs which originated not so far from North America. These terranes migrated towards the North American continental edge and became attached. The chapter discusses how this may have happened and how the terranes were modified after their arrival. The sum total of all these tectonic events is the complex geology of western North America.

In Chapter 12, "The Supercontinent Cycle," the authors hypothesize the existence of a cycle of continental assembly and fragmentation with a periodicity of 400 to 500 million years. Pangaea, the supercontinent that existed from 200 to 300 million years ago, is only the latest in a series. The existence in the geologic record of times of maximum orogenic activity going back to times early in earth history has been known since the 1950's. The authors propose that these orogenic maxima correspond to successive development of supercontinents, followed 100 million years or so later by continental breakup. They suggest that these cycles have had important bearing on the patterns of evolution and sea-level changes. They extend the hypothesis to suggest that the composition of the atmosphere and oceans may also depend upon the status of the continents in the pattern of breakup, reassembly and renewed breakup. Thus the very patterns of weather, climate and of life on earth may reflect the tectonic processes driven by the internal heat engine of the earth.

Right or wrong, this model of a supercontinent cycle offers a starting point from which to develop a holistic model of the processes, tectonic and otherwise, that have shaped the earth. The model offers less a final story than a challenge to continue our efforts to understand our planet.

Terranes

They are fault-bounded blocks of crust that accrete to the ancient cores of the continents. The process makes the continents increase in extent and reworks them into what amount to geologic collages.

• • •

David G. Howell
November, 1985

For more than a century geologists have been seeking to understand the engine that produces the great geologic features of the earth. The early investigators conceived of geosynclinal cycles: great crustal downbowings that fill with sediment, followed by upheavals that create young mountain ranges. That picture gave way to the theory of plate tectonics. In this view, put forward in the 1960's, the dominant directions of motion are horizontal: the brittle, outer layer of the earth consists of large crustal plates that are constantly shifting. Where plates move away from one another, rifts develop and new ocean basins form; where plates collide, chains of volcanoes erupt along lines parallel to the zone of the collision; where plates slide past each other, along faults such as the San Andreas fault in California, great earthquakes can occur.

It is becoming apparent, however, that a further revision is in order. The crustal patterns engendered by plate-tectonic activity are ephemeral: tectonic forces rework the original patterns, carving out fragments of crust, dispersing them and refashioning them into groupings of disparate crustal blocks. At the same time new crustal blocks arise from

volcanic processes and get swept into the reworking. The crustal plates are turning out, therefore, to be strange patchwork mixtures of crustal fragments, geologic collages consisting of pieces known as terranes.

The concept of terranes emerged in the 1970's, when conflicts over land use in Alaska required that the U.S. Geological Survey send teams of geologists to survey mineral resources. What they found was startling. The elucidation of a geologic pattern in one part of the state would lead to a prediction of

Figure 7.1 SIX TERRANES occupy the part of the coast of southeastern Alaska shown in this Landsat image. The coast is indented by Yakutat Bay; to its west the Malaspina Glacier descends from a group of peaks including Mount St. Elias and Mount Augusta. Under the ice and snow the rock represents parts of volcanic islands, parts of a displaced continental margin or the metamorphosed and remelted products of sediments within a sedimentary matrix. In each case the rock accreted to the ancient core of North America during the past 100 million years. In southern Alaska the terranes are elongated bodies resulting from the slicing of the crust by faults to the south. Even today the faulting continues to spread terranes northward.

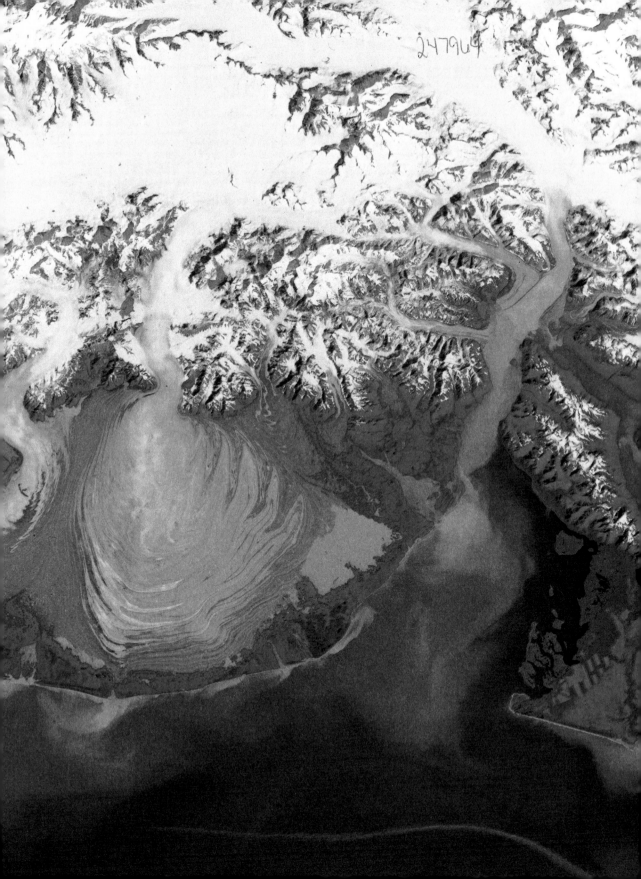

what the pattern should be a few tens of kilometers away. Yet the actual pattern would be markedly different: the rock would be the wrong age and have the wrong composition. In brief, the straightforward application of the plate-tectonic theory failed to account for the geology of Alaska. The entire state proved instead to be an agglomeration of crustal fragments. Alaska is the crustal flotsam and jetsam of the ancient, vanished ocean that preceded the Pacific. It is a collage of terranes dismembered and repositioned over the past 160 million years by the wanderings and collisions of crustal plates (see Figure 7.1). Pieces are still arriving from the south.

THE ADVENT OF PLATE TECTONICS

A brief review of the plate-tectonic theory helps to put terranes in perspective. After all, terranes represent only the newest aspect of the central effort in geology, an effort to grasp the vastness of time and comprehend the cumulative effects of slow movements in the earth. (Most tectonic processes plod along at rates equal to or less than the growth rate of a fingernail.) Fundamentally the crust of the earth has two domains: the oceanic crust, which is dense and homogeneous, and the continents, which are lighter and mineralogically heterogeneous. Terranes are the crustal fragments swept onto the ancient cores of the continents. In describing their histories one confronts the full consequences of the plate-tectonic theory, in which oceanic crust transports the continents as though it were a great, slow conveyor belt.

Oceans indeed are a central part of plate tectonics. Specifically, each ocean basin widens in the wake of diverging plates as magma, or molten rock, rises and solidifies to produce oceanic crust along a submarine ridge called an oceanic spreading center. Over geologic time the process may have slowed, and the network of spreading centers may well have decreased in extent. The present length of the network amounts to some 56,000 kilometers, and during the past two billion years the spreading velocities have probably averaged about five centimeters per year. The present-day Atlantic spreads at less than three centimeters per year; the most active parts of the East Pacific spread at about 16 centimeters per year. (These rates combine the velocities of the two plates moving in opposite directions from an oceanic spreading center.)

The multiplication of the two figures—the average spreading rate and the length of the spreading-center system—yields the estimate that new oceanic crust is formed today at a rate of 2.8 square kilometers per year. The area of the oceans is 310 million square kilometers. It follows that the oceans could have formed in a mere 110 million years. This is quite a realization. Before the theory of plate tectonics the oceans were taken to be the oldest parts of the earth, simply because they are the lowest parts of the surface. The idea was that old rock is colder, hence denser and lower, than young rock. The remarkable youth of the oceans is now well confirmed by the samples of oceanic crust amassed by the Deep Sea Drilling Project. In the earth today the age of the oceanic crust ranges from essentially zero along the submarine ridge crests that mark spreading centers to no more than 180 million years in the eastern Pacific, the part of an ocean floor most distant from a ridge. Over the past two billion years as many as 20 oceans may have been created and destroyed.

Oceanic crust rises at a ridge, moves across the width of an ocean basin and descends along a trench that marks what is called a subduction zone. The crust's surface, however, is not flat. On a broad scale the oceanic rock cools and compresses as it spreads away from the ridge crest where it was emplaced in the crust, and so the crust systematically sinks.

Moreover, mountains known as seamounts or oceanic plateaus are widely distributed across the ocean floor. Many are high enough to form islands. As I shall show, they can be swept up by plate-tectonic processes and thus become part of terranes. The primary constituent of a seamount is basalt, a dark volcanic rock that is rich in iron and magnesium and has a silica content of less than 50 percent. The rock ascends from "hot spots" under an oceanic plate. In fact, when the position of the hot spot is stable in the earth (that is, when the hot spot is stationary with respect to the core of the planet), long linear chains of volcanoes can form as the plate slides over the rising jet of magma. The Hawaiian Islands are part of such a chain (see Figure 2.6). The rate of growth of the chain is rather rapid as geologic processes go, and yet the hot spot itself is seldom more than a kilometer in diameter. The annual global contribution of oceanic basaltic piles to the augmentation of the continental crust is estimated to be only about .2 cubic kilometer per year.

SUBDUCTION-ZONE VOLCANISM

A greater contribution to the continental crust comes about when two plates collide. Along the zone of collision the denser plate descends, so that it encounters progressively greater temperature. The plate carries sediment and also water trapped in pores in the crustal rock. At a critical depth, usually between 100 and 150 kilometers, the water triggers a sequence of physical and chemical events, including the partial melting of rock, producing a magma that tends to be rich in volatile chemical elements, notably aluminum, potassium and sodium. The silica content ranges from 50 to 75 percent. The magma is stickier and more viscous than an oceanic, basaltic magma, and so it supports a buildup of pressure. The result is that the volcanism related to the earth's subduction zones tends to be explosive. Mount St. Helens and Krakatau are examples of the process, which has also formed numerous island arcs.

The plate-edge volcanoes are quite different from the intraplate islands and seamounts. For one thing the plate-edge volcanoes lie above great rising curtains of magma that parallel the trench marking the interface between two colliding plates. In all, the earth now has some 37,000 kilometers of plate-edge volcanic chains, and for each kilometer of such activity it is calculated that from 20 to 40 cubic kilometers of new siliceous material erupts in a million years. Worldwide, therefore, siliceous volcanic material joins the continental crust at a rate of roughly .75 to 1.5 cubic kilometers per year.

Clearly plate tectonics offers a rationale by which to examine terranes and classify additions of new material to the continental crust. In itself the oceanic crust is likely to contribute almost nothing, since the formation of oceanic crust at ridge-crest spreading centers is likely to be balanced by the loss of oceanic crust along subduction zones. The evidence is impressive. The assemblage of crustal rocks produced at spreading centers, called a MORB (midocean ridge basaltic) ophiolite and consisting of a characteristic sequence of three strata a total of six kilometers thick, is observed only rarely in folded mountain belts. (Mountain ranges almost always form by crustal folding.) Such belts are where MORB ophiolites would be found if oceanic crust augmented the continental masses.

Protruding above the oceanic crust, however, are the islands or island chains created by gentle, basaltic volcanism above hot spots and the island arcs created by the more explosive volcanism that parallels subduction zones. The ocean basins also include fragments of continental margins that broke away from the continents when rifting occurred and new oceans opened. Moreover, on the ocean floor there are blankets of sediments currently totaling 170 cubic kilometers worldwide. The sediments represent river-borne continental debris, planktonic fossil remains and chemical precipitates from the sea. Some of the material on the ocean floor is subducted. Nevertheless, much of the sediment, the basaltic seamounts, the volcanic island arcs and the continental fragments is destined to be swept together: it becomes terranes, which increase the size of the continents.

THE NATURE OF TERRANES

In geology the word terrain simply designates the lay of the land. In contrast, the term terrane (the full name is tectonostratigraphic terrane) designates a crustal block, not necessarily of uniform composition, bounded by faults. It is a geologic entity whose history is distinct from the histories of adjoining terranes. Terranes come in many sizes and shapes, and they have varying degrees of compositional complexity. India, for example, is a single great terrane. Some of its individual rock formations have ages exceeding a billion years; nevertheless, over the past 100 million years India has acted as a single mass. (It was part of the margin of the great, but now shattered, megacontinent Gondwana and then broke away and drifted north to a collision with the southern margin of Asia.) Conversely, the terranes that did not originate as a fragment of some earlier continent generally embody a fairly simple history spanning less than 200 million years, the normal maximum survival time for an ocean floor. The composition of such terranes tends to resemble that of a modern oceanic island or plateau. Some terranes consist chiefly of consolidated pebbles, sand and silt; they represent sedimentary fans that accumulated in an ocean basin, commonly between colliding crustal fragments.

The geometry of a terrane is the product of its history of movements and tectonic interactions. Terranes born on an oceanic plate retain their shape until they collide and accrete. Then they are subjected to crustal movements that modify their shape. For example, the terranes of the Brooks Range in Alaska are great sheets stacked one on top of another. Elsewhere in the cordillera, or chain of

a

b

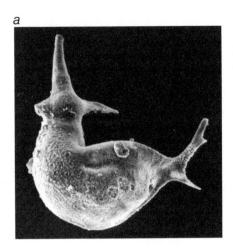

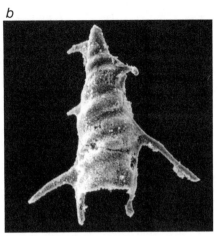

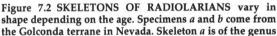

Figure 7.2 SKELETONS OF RADIOLARIANS vary in shape depending on the age. Specimens *a* and *b* come from the Golconda terrane in Nevada. Skeleton *a* is of the genus *Pseudoalbaillela*, some 290 million years old (enlarged 200 diameters); *b* is of the genus *Albaillella*, some 250 million years old (enlarged 300 diameters). Specimens *c* and *d* are

mountain ranges, in western North America the terranes are elongated bodies. The elongation reflects the slicing of the crust by a network of northwest-trending faults including the San Andreas fault of California. In Asia the terranes tend to have retained the shapes they inherited from episodes of rifting; some smaller terranes, however, were caught in collisions between the larger ones and got distorted. The assemblage of terranes in China is being stretched and displaced in east-west directions as India continues to squeeze Asia from the south.

The precise history of the movement of an individual terrane is not always known. Indeed, it is only recently that paths have been documented for a few of the earth's terranes. Since by definition terranes are fault-bounded and distinct from their geologic surroundings, each of them must have moved a distance at least equal to its longest dimension. The actual distances vary greatly. Some basaltic seamounts now accreted to the margin of Oregon have moved a minimal distance, from a nearby offshore origin. Yet similar rock formations around San Francisco have come as far as 4,000 kilometers across the Pacific. At a rate of just 10 centimeters per year a wandering terrane could complete a circuit of the globe in just 400 million years. Little wonder that the continents are patchwork agglomerations of terranes.

LINES OF EVIDENCE

How can the history of a terrane be reconstructed? Fundamentally the origin of each rock unit composing a terrane gives insight into the evolutionary history of the terrane. The sedimentary rocks indicate depositional environments of the past: they suggest ancient river gravels, coral banks, deltaic sands, muds of a continental shelf or muds of an ocean abyss. The age of the sedimentary rocks is also important. One aid in determining age is the fossil record. Until recently, however, the record was incomplete: the fossil evidence came mostly from rocks deposited in ancient shallow marine environments. Only in the past decade has it become possible to determine the age of rocks representing deeper oceanic deposits—rocks that are essential to an understanding of the events that created many of the earth's mountain belts. The breakthrough involved radiolarians and conodonts.

Radiolarians are single-cell organisms that appeared in the oceans as early as the Cambrian period, some 500 million years ago, and were abundant until as recently as 160 million years ago (see Figure 7.2). They occupied the upper levels of the ocean, but their skeleton consisted of silica, a substance of very low solubility at all levels of the sea. Accordingly abyssal muds are often rich in what is called radiolarian ooze: the accumulation of their

c

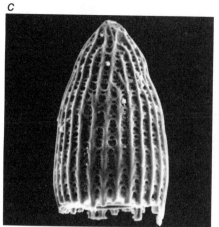

d

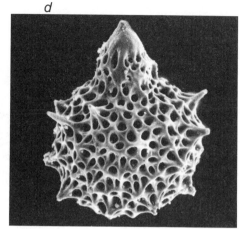

from the San Simeon terrane in southern California; *c* is of the genus *Archaeodictyomitra* and *d* is perhaps of the

genus *Stichocapsa*; both range in age from 100 to 150 million years and are enlarged 700 diameters.

skeletal silica. The rock called flinty chert, exploited for the making of arrowheads and knives by many peoples, is in fact the hardened ooze. Once the technique was perfected for extracting radiolarians from chert by dissolving the rock in a strong acid, thousands of chert assemblages could be dated.

Conodonts too are microscopic fossils, and like radiolarians they can now be extracted from rock by dissolving the rock in strong acids. Their biological identity was elusive. Now, however, it seems certain they are skeletal remains from the feeding ap-

paratus of an extinct group of small, marine, worm-like animals that lived from 570 to 200 million years ago (see Figure 7.3). Because conodonts are found in association with other fossil assemblages, it has been possible to establish a time scale of changing conodont morphology. In fact, for rocks more than 250 million years old the basis for radiolarian biostratigraphy (the determination of the age of a rock from the nature of the radiolarian fossils it holds) is the occasional joint occurrence of radiolarians and conodonts. By means of radiolarians and conodonts

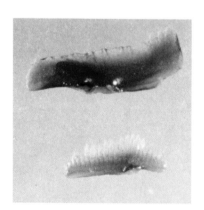

Figure 7.3 CONODONTS vary in shape and surface ornamentation depending upon their age. The specimens on the left are from the Brooks Range in Alaska and lived in shallow-water marine environments about 360 million years ago. The conodont in the center lived at shallow

depths about 325 million years ago; it was collected from a small fragment in a conglomerate in east-central Alaska. The specimens on the right, from Glacier Bay in southeastern Alaska, lived in deep water about 230 million years ago.

many rocks whose age was unknown a decade ago can now be dated. The results are often surprising. Sequences of old rocks are sometimes found to be lying on younger ones; hence great piles of strata seem to have been reshuffled. In some cases the stacking surfaces are parallel to the strata and the rocks show no obvious sign of repositioning.

Conodonts also are interesting in that they vary in color depending upon the maximum temperature attained by the rock containing them, an important matter in exploration for hydrocarbons and other minerals. Between 50 and 550 degrees Celsius, the organic matter in conodonts changes from pale yellow to brown, black, gray and white, and finally it loses all color (liquid hydrocarbons cannot survive at temperatures above 150 degrees). Thus the pale yellow to light brown conodonts on the left in Figure 7.3 indicate a temperature less than 90 degrees; the grey conodonts in the center indicate a temperature of at least 400 degrees, and the blue-black conodonts on the right indicate a temperature of at least 300 degrees.

Structural geology also has a part in the analysis of terranes. The reason is plain: the movement of rocks past (or over or under) one another imparts folds, crenelations, lineations and foliations, microscopic and macroscopic, all of which are a help in reconstructing the movement. Many of the folds one sees in the field, however, represent not the primary direction of motion but a later motion that consolidated terranes into tighter packages. The structural data are complemented by analysis of the composition of the rock. There seem to be no sure correlations between mineral assemblages or chemical composition and particular modes of origin, but there are generalizations that often hold true. For example, granite whose content of strontium is unusually rich in the isotope strontium 87 tends to result from the solidification of magma within an existing old continent, whereas granite poor in strontium 87 indicates an origin in an oceanic setting.

An important contribution to reconstructing the history of terranes derives from the study of remanent paleomagnetism: the alignment of minute magnetic particles in a rock, induced, at the time the rock formed, by the magnetic field of the earth. The investigations assume the earth's magnetic field is essentially that of a dipole, or bar magnet, coincident with the planet's axis of spin, so that the field lines at the Equator are horizontal (parallel to the surface of the earth) and the field lines at each pole are vertical (perpendicular to the surface). Between the Equator and the poles the field lines steepen. Most sedimentary and volcanic rock is deposited horizontally in sheets; hence the tilt of its remanent paleomagnetism can reveal the latitude at which the rock was formed. In addition the compass direction of the orientation of the magnetism can suggest that the rock has been rotated at some point in its geologic wanderings.

COLLAGE TECTONICS

Drawing on all these lines of evidence, I shall now take up, in terms of terranes, the rebuilding and reshaping of the continents, past and future. Some numbers are worth keeping in mind. The present global volume of continental crust is approximately 7.6 billion cubic kilometers; the oldest rock known is 3.8 billion years old. Dividing the first number by the second yields a simple linear estimate that the continents have grown at a worldwide rate of about two cubic kilometers per year — or roughly 65 cubic meters per second. The estimate may be too high; the growth processes in the hot, primitive earth may have been more rapid than the average. A host of crustal growth curves have been proposed; most of them hypothesize that between 70 and 80 percent of all crustal growth occurred more than two billion years ago. The final 20 to 30 percent of the current mass of the continents would then have accumulated over the past two billion years, at an average rate of from .7 to 1.1 cubic kilometers per year, a rate well within the rate of crustal contributions from modern volcanic arcs and oceanic seamounts.

The accumulation amounts to the plastering of terranes onto preexisting cratons, or continental nuclei, the oldest parts of the continental crust. The process can be followed in greatest detail through Phanerozoic time, the span of approximately 600 million years for which the fossil record of multicellular life is abundant. At the beginning of the span the existing continents (according to the paleomagnetic data) were isolated masses strung around the globe in the equatorial region. (The period from 700 to 500 million years ago was apparently an age of major continental breakup.) In the ensuing 350 million years the shifting of the continents resulted first in the agglomeration of two megacontinents, Gondwana and Laurasia, and then, 250 million years ago, in the union of the two to form the supercontinent Pangaea, a broadly crescent-shaped mass with a general north-south orientation (see Figure 7.4). The old continental nuclei, augmented by terranes that had accumulated since the beginning of the Phan-

Figure 7.4 ONE VAST OCEAN, Panthalassa, dominated the surface of the earth some 250 million years ago, when essentially all the planet's continental crust was agglomerated into a single supercontinent, Pangaea. Since that time Pangaea has broken into the continents of today and the oceanic crust of the Panthalassan basin has been completely subducted (directed downward back into the earth's mantle). Its place is now taken by the ocean basins of today. The crustal flotsam and jetsam of Panthalassa (consisting of fragments of continental crust along with crust created by volcanic activity) yielded terranes that now augment the continents ringing the Pacific. Such terranes appear in orange in Figures 7.5 and 7.6.

erozoic, then began to break up again some 200 million years ago along a new pattern of rifts resembling the modern 56,000 kilometers of globe-girdling oceanic spreading centers.

One can imagine, in the earth several hundred million years from now, the formation of a new supercontinent consisting of Asia and North and South America. The Pacific will have closed, following the subduction of the East Pacific spreading ridge. Meanwhile the Atlantic will have continued to widen. One can also predict that by then the colliding continents will have increased in size. The surface area of the continents ringing the Pacific today is 290 million square kilometers, of which the

newly accreted terranes (that is, the post-Pangaea material) contribute approximately 25 million square kilometers, or 9 percent. If one assumes the crust has an average thickness of 20 kilometers, the crustal growth rate in the Pacific over the past 200 million years has been 2.5 cubic kilometers per year. The rate is somewhat misleading: the tectonic collages ringing the Pacific include some large terranes consisting of displaced continental crust that formed before Pangaea broke up. Examples include the eastern half of Mexico, the Brooks Range of Alaska, parts of northeastern Russia and most of the Malay Peninsula. Still, the preliminary investigations do seem to indicate that growth rates for the continen-

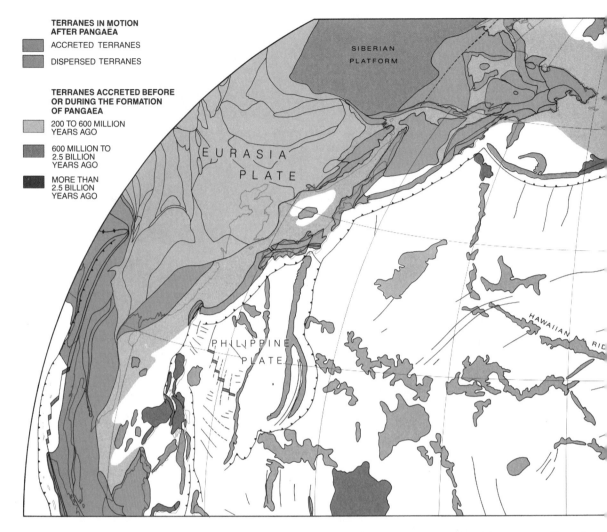

TERRANES IN MOTION AFTER PANGAEA

ACCRETED TERRANES

DISPERSED TERRANES

TERRANES ACCRETED BEFORE OR DURING THE FORMATION OF PANGAEA

200 TO 600 MILLION YEARS AGO

600 MILLION TO 2.5 BILLION YEARS AGO

MORE THAN 2.5 BILLION YEARS AGO

SIBERIAN PLATFORM

EURASIA PLATE

PHILIPPINE PLATE

HAWAIIAN RID

Figure 7.5 TERRANES OF THE NORTH PACIFIC on a map showing more than a fourth of the earth's surface. The Pacific plate dominates the figure. The colors indicate the ages of terrane accretion. The oldest terranes (*brown*) indicate the ancient cores of the continents. The youngest accretions (*orange*) make up approximately 9 percent of the

tal crust that rings the Pacific exceeded the global average rate of one cubic kilometer per year.

In this regard some recent studies of remanent paleomagnetism in limestone blocks of northern California are intriguing. The studies indicate that the limestone, which ranges in age from 85 to 100 million years, was deposited south of the Equator. Yet the age of sedimentary rock that now laps across the blocks suggests that the limestone (and associated basalts) became accreted to the margin of California no later than 38 million years ago. The figures require the plate carrying the limestone to

have moved northward at a rate of between 15 and 30 centimeters per year, which is faster than the plates are moving now. The figures lead me to speculate that the growth rate of the continents may also vary. Perhaps the growth comes in cycles hundreds of millions of years in duration.

THE IMPORTANCE OF SEDIMENTS

Sediments play far more than a passive role in the growth or diminution of the continents. They too become part of terranes. For one thing, thick piles of

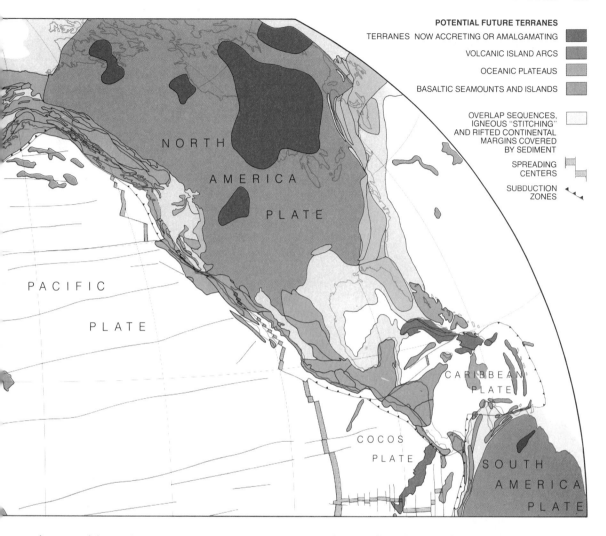

POTENTIAL FUTURE TERRANES

TERRANES NOW ACCRETING OR AMALGAMATING

VOLCANIC ISLAND ARCS

OCEANIC PLATEAUS

BASALTIC SEAMOUNTS AND ISLANDS

OVERLAP SEQUENCES, IGNEOUS "STITCHING" AND RIFTED CONTINENTAL MARGINS COVERED BY SEDIMENT

SPREADING CENTERS

SUBDUCTION ZONES

surface area of the continents around the Pacific and represent crustal debris swept up by the subduction of Panthalassa. Possible future terranes include hot-spot traces such as the Hawaiian ridge and its continuation, the Emperor Seamount chain, oceanic plateaus and volcanic island arcs over subduction zones.

sediments remain on the continents and accumulate along their rifted margins. Only about 30 percent of the sediments discharged from rivers make it beyond the continental margin and settle onto oceanic crust. In the second place, part of the sediments carried to subduction zones by the oceanic conveyor belt seems to get plastered onto the overriding crustal plate as an "accretionary prism" or becomes attached to the underside of the overriding plate. Then too, great piles of sediments are commonly caught between colliding crustal masses. An example is seen today in the Molucca Sea, where two island arcs are colliding. A sediment pile may be a terrane in itself; this is the case for the accretionary prisms in the region of Kodiak Island and the Gulf of Alaska. Alternatively, the pile may form the matrix in which terranes are embedded, as in the case of the Alaska Range.

In the world today the greatest single source of sediment is the towering landform resulting from the collision of India and Asia. There the Asian crust has overridden the Indian terrane, doubling the thickness of the crust and creating the Himalayan Mountains, and to their north the Tibetan plateau.

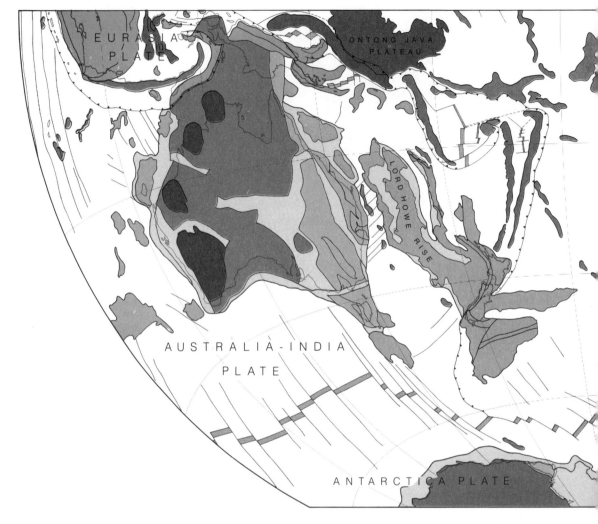

Figure 7.6 TERRANES OF THE SOUTH PACIFIC occupy a map that complements Figure 7.5. (Symbols are the same for both figures.) Some terrane amalgamation has already occurred: an island arc has collided with the northern margin of Australia, in New Guinea (see Figure 1.8); another island arc, the Vanuatu island has collided with the On-

Six great river systems, the Huang He, the Yangtze, the Irrawaddy, the Mekong, the Ganges-Brahmaputra and the Indus, drain the region, which amounts to only 4 percent of the world's land surface. Together they discharge into the oceans some 3.8 billion tons of sediment per year, or as much as 40 percent of all the sediment discharged by all the rivers of the earth.

The sediment is composed of silt and clay, along with rock and mineral grains. A further constituent is water, owing to the porosity of the solids, which is commonly as much as 50 percent. If the net den-

sity of the sediment is two grams per cubic centimeter, the volume of the sediment discharged each year from Asia's rivers is 1.7 cubic kilometers; the volume worldwide is from 4.5 to 6.8 cubic kilometers. As the sediment compacts and becomes stone the porosity decreases to almost zero. The worldwide discharge therefore amounts to some 3.3 to 4.9 cubic kilometers of rock per year. (I have assumed a rock density of 2.75 grams per cubic centimeter, which is a bit greater than the density of common quartz.)

The long-term fate of most of the rock is un-

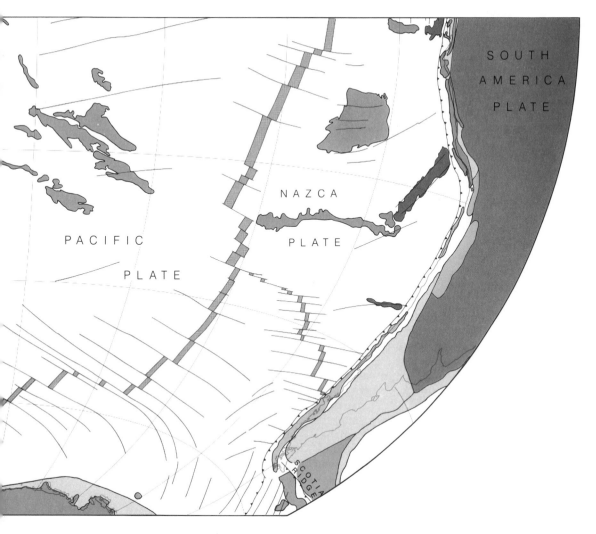

tong Java Oceanic plateau. The maps display results of
investigations of D. L. Jones, E. Scheibner, Z. Ben-Avraham, E. R. Schermer and D. G. Howell.

known. Some of it could be subducted; some could
be lifted between colliding continental massifs;
some could be torn away from its place of accumulation (say in a river delta or a deep-sea deposit) and
ultimately become accreted to a far-off continent.
Still, the 3.3 to 4.9 cubic kilometers of rock that can
form each year from sedimentation along continental margins and on the ocean floor far exceeds the .2
cubic kilometer contributed each year by basaltic
volcanism and the .75 to 1.5 cubic kilometers contributed by explosive subduction-zone volcanism.
Little wonder that sedimentary rock, or its metamorphosed equivalent, is a major component of

mountain fold belts. Indeed, the estimates I have
given suggest that as much as 75 percent of newly
formed continental crust could consist of sediment
and metamorphosed or melted products of sediment.

TERRANES OF THE NORTH PACIFIC

The making and shaping of continents out of terranes is best explored by examining specific regions
of the world. My own interest is the Pacific, which I
divide into quadrants. Each quadrant displays a distinctly different geologic pattern, reflecting con-

trasting histories of the accretion and dispersion of terranes (see Figure 7.5).

In the northeastern quadrant, where the Pacific washes the western coastline of North America, terranes consisting chiefly of island arcs and other oceanic material have piled onto the coastline throughout the past 180 million years. The cordillera inland from the coast includes a network of northwest-trending strike-slip faults. Here the motion is primarily horizontal, along the plane of the earth's surface; thus the newly accreted terranes have been distended into a sequence of slivers. Wrangellia, a well-studied terrane that once lay at, or even south of, the Equator, is a good example. Wrangellia crashed into Oregon some 70 million years ago. Subsequent faulting has spread fragments of Wrangellia northward, leaving parts in eastern Oregon, on the Vancouver and Queen Charlotte islands and throughout the Wrangell Mountains of southern Alaska. Central and western Oregon consists of terranes swept in after Wrangellia was in place.

The Brooks Range of northern Alaska has a quite different story. There huge, thin sheets of strata representative of a continental margin have been thrust over one another, transforming a geography that once measured at least 500 by 1,000 kilometers into a crustal stack 500 by 300 kilometers. Analysis of clues to the directions of flow of the sediment-bearing fluids that contributed to the strata indicates that the entire stack must have moved to its current position, but from where is a matter of debate. According to one hypothesis, the stack emerged by a counterclockwise rotation from the islands of the Canadian Arctic. The Canadian basin of the Arctic Ocean would then represent the depression left by the pivoting landmass.

The northwestern quadrant of the Pacific includes Asia, Japan and the Philippines. Here the continental crust consists of old continental fragments, each of them flanked, or even surrounded, by belts of terranes that accreted during the Paleozoic era, from 600 to 250 million years ago. In effect the Siberian platform seems to have been a backstop on which a succession of terranes built outward. Along the southern border of the platform volcanic arcs and other crustal chunks piled up in the early Paleozoic, from 600 to 400 million years ago; they formed the Baikalean fold belt, the mountainous region between Mongolia and the Sea of Okhotsk. Then, from about 300 million years ago to about 60 million years ago, when India came crashing in, a number of terranes (Tarim, Yangtze, the Sino-Korean massif, Indochina and finally India) came together to form Asia.

A similar episode of accretion is now in progress. The Ontong Java oceanic plateau is probably a rifted fragment of continental crust. Today it is mostly submerged. In size it is comparable to the Yangtze terrane of Asia. Accreted against the southern side of the Ontong Java plateau is part of the New Hebrides volcanic arc. Many other arcs lie nearby. If the Coral Sea closes up—an event that is hard to predict, since in some places the sea is opening, while in others it is closing—the Ontong Java plateau, wreathed with accreted volcanic arcs, is likely to form a major addition to Australia.

TERRANES OF THE SOUTH PACIFIC

The southwestern quadrant of the Pacific, including Antarctica, Australia and New Zealand, is a region characterized tectonically by a radial dispersion: the landmasses result from the breakup of part of Gondwana that began between 120 and 100 million years ago, when a three-pronged rift system developed (see Figure 7.6). One prong created the Tasman Sea; the other two prongs separated Antarctica from Australia and the Campbell Plateau of New Zealand. Fold belts in western Antarctica, eastern Australia and New Zealand suggest the earlier history of the region. Evidently episodes of accretion had built the continental crust outward from nuclei now positioned in eastern Antarctica and western Australia.

That leaves the southeastern quadrant of the Pacific, which includes the western shore of South America. It consists of contrasting regions. In the southern part of the quadrant new terranes are being carved in the Scotia Sea, which is cutting between the southern Andes and the Antarctic Peninsula, leaving small, fault-bounded crustal blocks such as South Georgia and the Orkney Islands. Northward, from southern Chile to Peru, the Andes extend in a nearly straight line for 3,000 kilometers. The region beckons for more study, but from what is now known there appears to be little sign of terrane accretion, in spite of the past 200 million years of subduction. (Oceanic crust is being subducted under the continental crust of Chile and Peru.)

From central Peru to the Caribbean, western South America displays a patchwork fabric of accreted terranes. Here, as at the southern end of

South America, crustal dispersion is active. For example, the petroleum-rich sediments under Lake Maracaibo in northwestern Venezuela fill part of a sag in the crust left when the Antilles island arc of the Caribbean plate slid eastward past the northern edge of South America. Like two bulldozers, the Scotia arc in the south and the Antilles arc in the north are advancing into the Atlantic, and along the flanks of the advance lies crustal wreckage, which presumably will serve as raw material for the building of future terranes.

The concept of terranes is becoming part of the plate-tectonic theory as the patterns and processes of continental growth are examined. Overall the budget for continental growth is dynamic. Volcanoes contribute from .75 to 1.5 cubic kilometers per year, and hot spots contribute .2 cubic kilometer per year. Meanwhile, owing to erosion, the continents lose as much as four cubic kilometers per year, but as much as three-fourths of the loss is recycled: the erosional sediment is lifted, folded, sometimes metamorphosed and sometimes melted, in the collisional, accretionary processes that raise mountains along the continental margins. The averaging of billions of years of geologic history probably masks pulses or even cycles of continental growth. In the aftermath of the breakup of Pangaea 200 million years ago an entire global-scale ocean, Panthalassa, has been consumed and the circum-Pacific continental crust has grown at a rate of as much as 2.5 kilometers per year, which seems to have exceeded the average rate of long-term growth. The events in the Pacific seem, therefore, to require complementary periods of tectonic tranquillity. Charting the detailed history of the earth in terms of terranes is the challenge now facing geology.

Ophiolites

The crust of the earth under the oceans is different from the crust of the continents.
Ophiolites seem to be fragments of oceanic crust on land. They are thus clues to
how oceanic crust forms and spreads.

• • •

Ian G. Gass
August, 1982

Plate-tectonic theory proposes that oceanic crust, which forms some 70 percent of the earth's solid surface, has been and is being constantly created at the axes of oceanic ridges and rises. Thereafter, by the processes of sea-floor spreading, it moves away from these axes and ultimately plunges into the earth's interior along subduction zones. Since the oceanic crust has this built-in self-destruct system, no part of the present oceanic crust is much more than 200 million years old. In contrast, the continents, being lighter than oceanic crust, are not easily subducted and move passively over the face of the earth in response to sea-floor spreading and plate-tectonic processes. Indeed, continental rocks preserve evidence of earth history going back almost four billion years.

Occasionally a fragment of oceanic crust, instead of being subducted, is preserved at the leading edge of a plate that rides over a subduction zone. To describe this process Robert G. Coleman of Stanford University has coined the term obduction: the reverse of subduction. Coleman has also calculated that less than .001 percent of all oceanic crust has been obducted and remains on dry land. Small though these remnants are, they provide unique

information on the processes currently operating under the axes of oceanic ridges and rises. They also yield clues to the evolution of ancient oceans, the mechanisms of collisions between plates and the position of ancient destructive plate margins, and they strongly suggest that plate-tectonic processes have been active for at least the past billion years. These on-land fragments of oceanic crust are known as ophiolites (see Figure 8.1).

Like many other geological terms "ophiolite" has since the advent of plate tectonics taken on a new

Figure 8.1 OPHIOLITE ON THE ARABIAN PENISULA is represented by the dark-colored mountains running from the upper left to the lower right in this Landsat picture of northern Oman. The body of water at the upper right is the Gulf of Oman, which connects the Persian Gulf and the Indian Ocean. The mountainous area is known as the Samail nappe. Light-colored area to the left of the dark-colored ophiolite is continental rocks, mainly limestones; they are underlain by granitic rocks of the Arabian continental plate. The dark color of the ophiolite results from the abundance of basalts, gabbros and peridotites. The small patches of red along the coast and in other areas are vegetation. The field of view in the picture is 130 kilometers across.

meaning and significance. Even before this latest revolution in earth science, however, the meaning of the term evolved with changes in the understanding of geological processes. The term first appeared in the geological literature in the 1820's, when Alexandre Brongniart of France coined it to describe rocks also known as serpentinite or serpentinized peridotite: a type of igneous rock usually found in areas deformed by tectonic processes. "Ophiolite" is derived from the Greek *ophis*, meaning snake or serpent; therefore it has the same meaning as serpentinite. Both terms are appropriate only in that both the rocks and some reptiles have a mottled green appearance. Other than demonstrating the addiction of European geologists to the classical languages, "ophiolite" had little significance and was used in an ad hoc way throughout the 19th and early 20th centuries to describe serpentinized peridotites and the rocks commonly associated with them.

In the 20th century there were two major changes in the usage of the term. In 1906 Gustav Steinmann of Germany noted the close association of serpentinized peridotite with other igneous rocks and deep-water sediments (such as radiolarite) in the Alpine fold mountains around the Mediterranean. Later, in honor of this outstanding geologist, the association of serpentinite, radiolarite and pillow lavas (lavas erupted under water) became known as the Steinmann Trinity. Then, through common usage, "Steinmann Trinity" and "ophiolite" became synonymous. In other words, "ophiolite" no longer referred to one kind of rock but to an assemblage of related rocks.

Steinmann particularly emphasized the association between deep-water sediments and the serpentinite and the pillow lavas. Others developed this lead and proposed that ophiolites were masses of igneous rock emplaced in geosynclines: huge linear depressions in the earth's crust that become filled with sediments. Some workers believed the igneous rocks were intruded into the layers of sedimentary rock as sills (horizontal intrusions); others saw them as immense balloons of magma (molten rock) erupted onto the surface of the sediments along the flanks of the geosyncline.

It was visualized that once the balloon had erupted its skin chilled and fractured and was invaded by dikes (vertical intrusions) from the balloon's still-molten interior. Thereafter pillow lavas erupted onto its surface. As the molten interior crystallized,

the heavier minerals settled to produce layered peridotites and overlying layered gabbros.

Here I must pause briefly to define what is meant by terms such as "peridotite" and "gabbro." In this context they apply to the rocks of the oceanic crust and the underlying uppermost part of the earth's mantle. The uppermost part of the mantle is thought to be formed of peridotite, a rock that consists almost entirely of the magnesian minerals olivine $[(Mg,Fe)_2SiO_4]$ and pyroxene $[Ca(Mg,Fe)Si_2O_6]$. Also present, however, is dunite, a rock that consists almost entirely of olivine.

The oceanic crust is formed of rocks, such as basalt and gabbro, that are somewhat richer in silica (silicon oxides). Basalts are fine-grained; gabbros, having crystallized more slowly, are coarse-grained. Both, however, consist of the same minerals: olivine, pyroxene and plagioclase $(NaAlSi_3O_8$—$CaAl_2Si_2O_8)$. Basalts form the upper one to 2.5 kilometers of the oceanic crust (exclusive of the overlying sediments), gabbros the lower 3.5 to 6 kilometers.

To return to the geosyncline model of ophiolites, it developed in the 1930's, 1940's and early 1950's and remained the consensus among geologists until the mid-1960's. The essence of the model is that ophiolites were interpreted as being the result of magmatism during the initial stage in geosyncline development. They were therefore in situ (autochthonous) igneous rocks and, whether they were intruded as sills or erupted, they were interleaved with the sedimentary rocks of the geosyncline. It is this basic concept that has changed. Today almost all ophiolite masses are regarded as being allochthonous, that is, they were formed elsewhere and were transported tectonically to their present position.

The model now widely accepted is that an ophiolite is a fragment of oceanic crust formed at the axis of an oceanic ridge or rise, moved across the ocean floor by sea-floor spreading and finally lifted above sea level. Concurrently with this switch from an in situ to a transported origin for ophiolites the concept of the geosyncline, which once dominated the geological literature, has been virtually abandoned.

The complete change in ideas on the origin and significance of ophiolite complexes was brought about by a variety of investigations. First it was demonstrated that virtually all ophiolites are allochthonous, having been brought in contact with the adjacent rock formations by tectonic processes. Second, detailed studies of ophiolites in the eastern Mediterranean, particularly the Troodos massif on the island of Cyprus, showed that their internal structure was not compatible with their having been formed in situ in a geosyncline; it could be realistically explained only on the basis of magmatic processes operating at oceanic ridges and rises, where new oceanic crust is generated. Third, it was noted that in each of the eastern Mediterranean ophiolites the same rock sequence could be recognized and could be compared with layers in the sea floor deduced from geophysical evidence. Fourth, it was shown that ophiolite rock types were similar to those of rocks that had been dredged from the deep-sea floor.

The similarity of ophiolite sequences to one another and to oceanic sequences and the comparability of oceanic and ophiolitic rocks (demonstrated by Nikolas I. Christensen of Purdue University and Matthew H. Salisbury, now at Dalhousie University) led to the general acceptance of ophiolites as fragments of oceanic crust formed at oceanic ridges and rises. Since then it has been common practice to use ophiolite data to clothe the necessarily meager skeleton of oceanographic data on oceanic-ridge-and-rise structures. This view became widely accepted in the 1970's, but even so when those attending the Geological Society of America's Penrose Conference on ophiolites undertook to redefine the term in 1972, they stressed that it should be applied only to a particular rock association and should not have any connotations of origin.

A complete ophiolite, such as the relatively undeformed ophiolite of the Troodos massif of Cyprus, is an orderly sequence of rock types. In most instances the processes of obduction have disrupted the originally coherent mass and have spread parts of the ophiolite over a large area. The component parts of such a dismembered ophiolite can often be fitted back together by unraveling the tectonic jigsaw puzzle. It is easier, however, to study relatively undeformed ophiolites such as the Troodos massif and the Samail nappe in the Sultanate of Oman on the Arabian peninsula. Therefore it is on these studies that much of the following generalized description is based (see Figure 8.2).

Starting from the top and going down, most ophiolites are overlain by marine sediments. In some instances, as with the Cyprus and Oman ophiolites, the sediments are deep-water ferromanganoan mudstones or radiolarites. In others the sediments are similar to those on continental shelves or adjacent to arcs of volcanic islands. Therefore they indicate that the oceanic crust, from which the

ophiolite was derived, must have been near the margin of a continent or of an island arc at some stage of its history. More than anything else these sediments provide evidence on the environment of the ophiolite within an ocean basin before it was lifted above sea level; in this regard studies such as those of Alistair Robertson of the University of Edinburgh and Alan Gilbert Smith of the University of Cambridge have proved particularly valuable.

The top of the ophiolite proper consists of extruded basalts. Many of these rocks have the pillow forms characteristic of lavas that have erupted under water. Indeed, such lavas have been filmed in the act of formation by divers during a submarine eruption off Hawaii. Pillow lavas have also been repeatedly identified by manned and unmanned submersibles investigating the floor of the axial rift of the Mid-Atlantic Ridge and the axial zone of the East Pacific Rise.

Some lava pillows are nearly spherical; others are more elongated and might better be termed bolsters (see Figure 8.3). The form of the pillows probably depends on the configuration of the sea floor onto which the lava is erupted. On steep slopes the melt forms globules that roll down the slope and accumulate at the base. The elongated pillows probably form on gentler slopes, where the lava flows downslope for some distance before solidifying. When the lava is erupted in hollows or on flat surfaces, it is unlikely to be pillowed, and unstructured layers of varying thickness seem to form.

Pillow lavas are known to form in very deep water, in shallow water and even where lavas erupted on land flow into water. How much water was above them at the time of their formation is indicated by their content of gas vesicles, or bubbles. Such vesicles are formed when the gas dissolved in a molten rock separates from the melt as the pressure is lowered. If the pressure of the overlying water is sufficiently high, however, the gas does not separate and no vesicles form. By measuring the vesicularity of basalts from known depths James G. Moore of the U.S. Geological Survey has shown that there is a crude correlation between the depth below the surface of the water at which a lava was erupted and the vesicularity of the lava; with increasing depth (pressure) the vesicularity decreases. This can give only an approximate estimate of the depth because the abundance of the gas in the magma varies. Moreover, sea-floor spreading can move a lava to a depth other than the one at which it was erupted. Still, vesicularity is a further

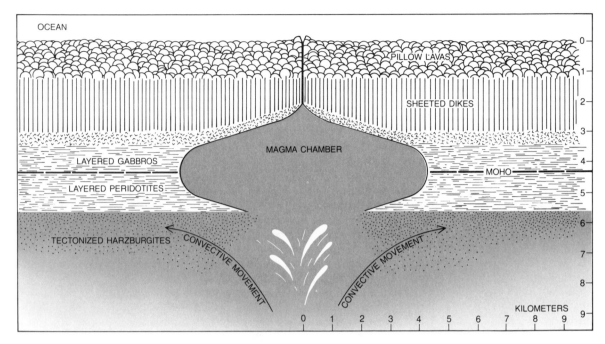

Figure 8.2 SCHEMATIC CROSS SECTION OF OCEANIC CRUST at a spreading center is based upon ophiolite studies. Sediments overlying the crust are not shown. The figure shows the rise of magma to a central magma chamber during convective movement in the mantle. Compare this figure with Figure 2.1, which shows a highly faulted oceanic crustal sequence. The two figures show the possible range in structure of the oceanic crust formed at different spreading centers. Also compare the size-postulated magma chamber with the considerably smaller one shown in Figure 2.1.

clue to the origin of ophiolites at oceanic ridges and rises.

Most of the lavas in ophiolites are basalts, similar to those dredged from present-day oceanic ridges. In Oman, however, it is only the lower part of the pile of lavas and the dikes under it that resemble the oceanic-ridge basalts in chemical composition. The upper part of the lava sequence differs geochemically from the lower, and Julain A. Pearce and Tony Alabaster of the Open University in England have convincingly demonstrated that these upper lavas were erupted in an island-arc environment later than the main event of sea-floor spreading. In Cyprus all the lavas have a composition comparable to that of volcanic rock erupted above a subduction zone.

The top of the lava pile consists entirely of extruded rocks. Farther down dikes become increasingly numerous. Many of the dikes are sinuous and seemingly had to wriggle their way upward through the lava pile. The thickness of the lava pile varies, but in the Troodos massif it is about one kilometer and in Oman it ranges between .5 kilometer and 1.5 kilometers. Near the base of the sequence the ratio of dikes to lavas is about 1 : 1; then, within a downward distance of 50 to 100 meters, the abundance of dikes increases from 50 to 100 percent and there is little or no lava between them. Moreover, the disposition of the dikes is not sinuous but a regular and often almost vertical array. Indeed, this part of the ophiolite has been likened to a pack of cards standing on edge. It has been termed the sheeted-dike complex (see Figure 8.4).

In the 1950's, before the advent of plate tectonics, it was the sheeted-dike complex that presented the main conceptual problem in ophiolite studies. Nowhere else had a rock complex consisting entirely of dikes been found. In classic fossil volcanic areas such as the Hebridean province of northwestern Scotland the abundance of dikes with respect to the total outcrop is less than 10 percent. Therefore the search was always on for the host rock in which the dikes had been emplaced. Any rock that looked slightly different was studied in detail; the commonest contenders as host to the dikes were structureless flows of basalt. No one, as far as I am

Figure 8.3 PILLOW LAVAS are seen in the Wadi Jizzi of the Samail nappe in Oman. These elongated pillows, which are more like bolsters, were probably erupted over a gentle slope on the ocean bottom and flowed for a short distance from the top right to the bottom left before solidifying.

aware, then thought of the possibility that there was no host rock at all. Only after the 100 percent dike structure of the sheeted-dike complex had been demonstrated in the undeformed and well-exposed Troodos massif was it sought and identified in more deformed masses.

When the plate-tectonic model was proposed, it was quickly realized that the axis of an oceanic ridge or rise was just the place where a 100 percent dike complex would form. The current concept is that most of the dikes are injected along a narrow zone, no more than 50 meters wide, at the axis of the ridge or rise and that the dike material is moved away from the axis by sea-floor spreading. Recent studies of sheeted-dike complexes, unencumbered by preconceived geological ideas, show that this part of an ophiolite complex does consist of 100 percent dike.

Johnson R. Cann and Rupert G. W. Kidd, working at the University of East Anglia, identified within the dikes the phenomenon of one-way chilling, which can be explained as follows. Magma being injected into a fissure in cold rock will form a dike-like body. The melt will cool most rapidly, and will therefore form the finest-grained rock, where it is in contact with the cold host. Cann and Kidd recognized that in the Troodos massif many dikes, instead of having two chilled margins, have only one. In addition, more of the chilled margins were on one side of the dikes than the other; this is the phenomenon of one-way chilling.

Cann and Kidd proposed that at an oceanic ridge magma would tend to be injected along the still liquid or softer axis of an earlier dike and so would split the two halves of the earlier dike, with one half moving in one direction and the other half in the

Figure 8.4 SHEETED-DIKE COMPLEX is seen in the Samail nappe. Each dike, originally an intrusion of molten rock, is a vertical sheet of rock. Each was intruded upward along the axis of an oceanic ridge or rise. The outward convective movement of the underlying mantle (see Figure 8.2) produces tension, and when a tension crack opens, magma from the chamber escapes upward and another dike is formed. The dikes are on the average one meter thick. Therefore the sea floor spreads in one-meter jumps every 50 to 100 years rather than in the smooth one to two centimeters per year visualized in the models of theoretical geophysics.

opposite direction. On each side of the ridge the dikes would show a preferential one-way chilling on the side farther from the ridge axis. On this basis it has been inferred that the oceanic ridge where the north-south dikes of the Troodos massif formed lay to the west of the present-day outcrop. It should be emphasized that the statistical excess of one-way chills in one direction over those in the other is small and also that continuous outcrops with a statistically valid number of dikes are found only rarely. One-way chilling is an attractive concept and one that seems intuitively acceptable, but it is by no means proved.

Although the sheeted-dike complexes are the best evidence that ophiolites formed along zones of tension in oceanic ridges and rises, such complexes are not always present. For example, in the Voúrinos ophiolite in Greece, as it has been described by Eldridge M. Moores of the University of California at Davis, sheeted dikes are poorly developed. It has been suggested that sheeted-dike complexes form only when sea-floor spreading is slow enough to allow a dike to solidify before the next one is injected. Since oceanic crust cools rapidly and seawater is percolating through all this material, the explanation is questionable. Moreover, the Oman ophiolite, the fossil of an oceanic ridge where the sea floor spread in both directions at the high rate of two centimeters per year, has a superb sheeted-dike complex. Although the absence of sheeted-dike complexes from some ophiolites remains a geological problem, it is worth noting that many ophiolites originally described as being without a sheeted-dike complex have been found on closer examination to have one.

The dikes of a sheeted complex must have come

from an underlying source of magma. It is therefore not surprising to find that with depth the basaltic dikes give way, over a distance of between 10 and 100 meters, to gabbroic rocks that have a similar composition but a markedly coarser-grained texture. Detailed mapping of these plutonic complexes (igneous rocks formed well below the surface) suggests that the uppermost gabbros, which form a layer between 10 and 300 meters thick, were formed by the melt cooling and crystallizing against the roof of a chamber from which the magma flowed toward the surface. Below this layer the gabbros and the underlying peridotites, if there are any, are markedly layered. Until recently it had been widely accepted that the layering was the result of the minerals' crystallizing out of the melt and then settling to the bottom of the magma chamber.

Alexander R. McBirney and Richard M. Noyes of the University of Oregon have proposed an alternative mechanism as a result of studying the classic layered gabbros of Skaergaard in eastern Greenland. They suggest that gradients of chemical composition and heat cause minerals to crystallize out of the melt along horizontal planes within it; no movement of the crystals is needed to account for the layering. Whether the crystals settle or form layers in place, however, the inference of the layered plutonic rocks is that there is a large body of magma below the surface along the axis of an oceanic ridge or rise. Work on the Troodos massif by Cameron R. Allen of the University of Cambridge has suggested that along this slow-spreading (one centimeter per year) axis there were numerous magma chambers four or five kilometers in diameter. For the faster-spreading (two centimeters per year) Oman ophiolite the magma chambers seem to have been some 20 kilometers in diameter.

The evidence of these layered plutonic rocks is that below the axis of the oceanic ridges there were magma chambers whose dimensions depended essentially on the input of heat into the ridge from the underlying mantle of the earth and on the rate of cooling brought about by sea-floor spreading. But do these layered rocks represent a single body of magma crystallizing completely or, as seems intuitively more likely, was the magma body fed periodically from below? Studies by E. Dale Jackson of the U.S. Geological Survey on the Voúrinos mass in Greece, by Cameron Allen on the Troodos massif and by John D. Smewing and Paul Browning of the Open University on the Samail nappe in Oman show that in all these ophiolites the

melt was periodically replenished by new batches of magma.

These workers, analyzing separate ophiolites, have shown that the layered sequences consist of repetitive cycles starting with dunite and giving way upward to peridotites and then to gabbros. The cycle starts again and is repeated many times throughout the layered sequence. The inescapable conclusion is that the composition of the melt in the main chamber was reset periodically by a fresh influx of magma.

At the base of all complete ophiolite sequences is a tectonized (deformed) peridotite consisting almost entirely of the minerals olivine and orthopyroxene [$(Mg,Fe)SiO_3$]. Peridotites of this composition are called harzburgites, and so they are described as tectonized harzburgites. The chemical and mineralogical homogeneity of the tectonized harzburgites suggests that they represent material of the uppermost part of the mantle from which basaltic liquid has been extracted. This proposal is supported both by the presence within the harzburgites of gabbro pods that are best explained as batches of basaltic melt that crystallized before they could escape from the mantle and by localized patches of peridotite of a different composition (lherzolites) that probably represent remnants of mantle from which little or no basaltic magma had been extracted.

Most workers accept that the tectonized harzburgite represents depleted mantle, the residue from which basaltic melts have been extracted to form the overlying rocks: from the bottom up the layered plutonic rocks, the sheeted-dike complex and the sequences of lavas. The model proposed to explain these features is that at depths of 25 or 30 kilometers basaltic magma first separates from the mantle. The two components, the magma and the harzburgite, move upward under the influence of convection in the mantle.

The rising magma collects into balloonlike bodies called diapirs, one kilometer to five kilometers in diameter. The diapirs move upward with the ascending mantle material, and as they do so olivine crystallizes out of the melt to form a crystal layer at the bottom of the diapir. The melt escapes into the main magma chamber but the olivine remains in the mantle as lenses of dunite within the tectonized harzburgite. Although the harzburgite is hot, it remains solid. Therefore it is deformed as it moves upward and then outward under the axis of the oceanic ridge or rise.

The available evidence suggests that the newly formed oceanic crust is transported by this convective movement in the underlying mantle. Detailed investigations, particularly those of Adolphe Nicolas and his colleagues at the University of Nantes, support that view. They also demonstrate that the deformation took place at about 1,000 degrees Celsius and imposed a linear fabric, perpendicular to the orientation of the overlying ridge axis, on the tectonized harzburgite.

Ophiolite studies have thus revealed much about the structural and magmatic features at fossil oceanic ridges and rises. What else do they reveal about the oceanic crust? Oceanic ridges and rises are repeatedly (on the average of every 30 kilometers) offset by fractures. These fratures, known as transform faults, make it geometrically possible for rigid plates to move over the face of a nearly spherical earth. Even if ophiolites represent only .001 percent of the subducted oceanic crust, one of them, if it preserves more than 30 kilometers of oceanic-ridge, is also likely to preserve a transform fault. Indeed, transform-fault structures have been investigated on Cyprus by my former colleague Kapo Simonian, on Masirah Island in Oman by Ian Abbott and Frank Moseley of the University of Birmingham and in western Newfoundland by

John F. Dewey and his colleagues at the State University of New York at Albany. The features displayed by these structures at various levels revealed by erosion enhance the understanding of present-day transform faults.

Another feature of ophiolites also relates them to oceanic crust: their metamorphism, that is, the fact their rocks have been greatly altered since they originally formed. Virtually all igneous rocks dredged or drilled from oceanic crust away from the axes of oceanic ridges and rises have been metamorphosed. These oceanic metamorphic rocks, unlike most continental metamorphic rocks, show no directional fabric. Therefore they were altered without the deformation that commonly accompanies continental metamorphism.

Since the metamorphic processes operated while the ophiolites were part of the oceanic crust, and since there are no directional fabrics, the main agent of metamorphism must have been heat. The main source of heat is the underlying magma. The metamorphic processes involved the circulation of seawater through newly formed, still-hot oceanic crust with a thermal gradient in excess of 150 degrees C. per vertical kilometer (see Figure 8.5). The water is believed to circulate by convection in a single-pass

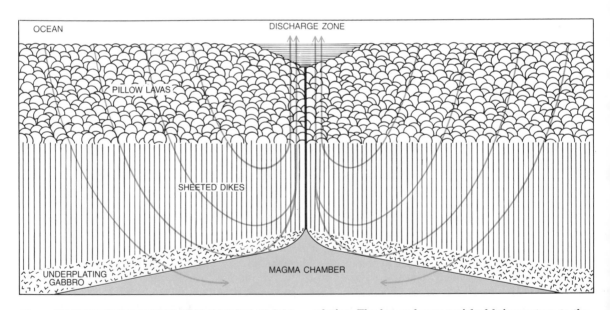

Figure 8.5 SEAWATER INFILTRATES HOT SEA FLOOR in the immediate vicinity of the spreading axis. The water is heated as it moves through the newly created oceanic crust and incorporates those elements that go readily into solution. The hot and now enriched brines return to the surface of the ocean bottom along fracture zones. The dissolved elements are precipitated out where the hot water reenters the ocean.

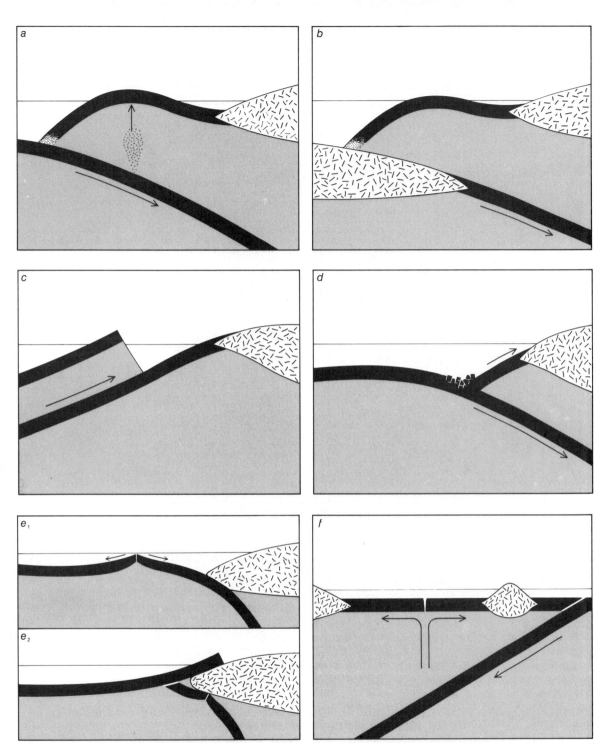

Figure 8.6 EMPLACEMENT OF OPHIOLITES on continental margins has been ascribed to several mechanisms. The problem is how to get oceanic crust and mantle up more than 5 or even 10 kilometers over continental crust. Proposed mechanisms include: (*a*), serpentinization of mantle above a subduction zone; (*b*) and (*c*), collision of a continental margin with a thrust fault that extends into the mantle and thus represents an incipient or more mature subduction zone; (*d*), formation of fragments in a subduction zone; (*e*), obduction of thin crust during subduction of a ridge, and (*f*), emplacement from a back-arc region. The collision mechanism (*b* and *c*) seems most likely to produce most large ophiolite complexes, possibly from a back-arc environment.

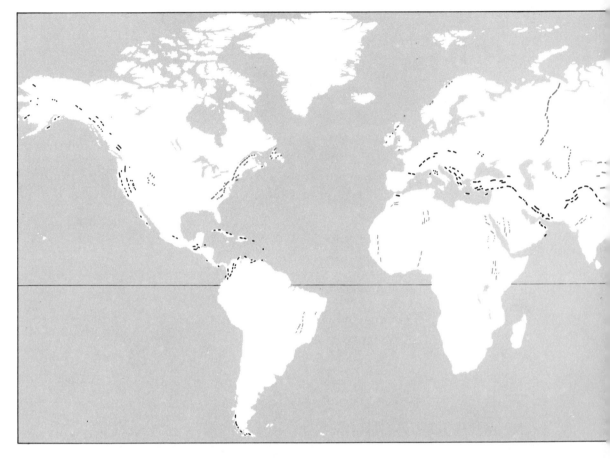

Figure 8.7 WORLDWIDE DISTRIBUTION OF OPHIO-LITES. The lines in black are ophiolites less than 200 million years old. The lines in color are ophiolites between 200 and 540 million years old. The lines in gray are ophiolites between 540 million and 1.2 billion years old. The younger ophiolites are those related to the present cycle of plate tectomics. They include the ophiolites emplaced around the Pacific, and all are close to the sites where oceanic crest is being subducted. The next-oldest ophiolites (running along the Appalachians, north into Nova Scotia

cycle, with the permeable rocks of the oceanic crust being recharged with water continuously over a wide area and discharging water above thermal highs.

The convecting seawater leaches metals out of the oceanic crust and also redistributes silicon and other elements. In returning to the surface the metal-enriched fluids are channeled along faults, and so the emission of the fluids into seawater is localized. In favorable sheltered locations, such as depressions in the sea floor, chemical reactions between the hot, metal-enriched brines and the seawater lead to the precipitation of sulfides of iron and copper and the formation of massive bodies of sulfide ores, and also to the precipitation of iron with manganese to form ferromanganoan sediments. Hydrothermal

vents emitting metal-rich brines have now been observed along the East Pacific Rise from manned submersibles.

Interesting developments in the understanding of plate-tectonic and related processes have resulted from these metamorphic studies. For example, it has long been held, particularly by Nikolas Christensen and Matthew Salisbury, that the horizontal layering in the oceanic crust, identified through the velocity of earthquake waves, is a metamorphic phenomenon and that the change from the velocity characteristic of the second layer from the top (5.07 kilometers per second) to the velocity characteristic of the layer below it (6.69 kilometers per second) represents the change from one type of metamorphism to another. Certainly in ophiolites the change

and Newfoundland and continuing into Ireland, Scotland and Norway) mark the closing of the Iapetus Ocean in the Paleozoic era. The ophiolites in the U.S.S.R. also mark a Paleozoic plate-collision suture.

from the dike complex to gabbros occurs at about the same level as a change in the type of metamorphism. Both changes could influence the earthquake-wave velocities of the oceanic crust.

So far we have been concerned with the structure, composition and metamorphic features of ophiolites and what they tell us about oceanic geology. There is a separate question raised by ophiolites. If they represent oceanic crust formed at oceanic ridges and rises, how are they emplaced near the destructive plate margins adjacent to continents and island arcs? It is to this process that the term obduction is applied. The actual processes of obduction are poorly understood, and further considera-

tion of the term and its implications is needed (see Figure 8.6).

Almost as soon as the processes of plate tectonics were proposed the term subduction came into the geological literature. The word is from the Latin *sub*, under, and *ducere*, to lead. Hence "subduct" literally means "to lead under." "Ob-" rather than "sub-" implies that the movement is in a direction or manner contrary to the usual one. Therefore obduction implies a movement contrary to subduction, a movement over rather than under. It also implies an upward movement of the oceanic crust, and just how this movement comes about is a puzzle.

The simplest explanation of the onland presence of ophiolites is to regard them as fragments of oceanic crust attached to a continent or to an island arc. As a result of the subduction of oceanic crust seaward of the eventual ophiolite, both the continent or island arc and the oceanic crust attached to it are underlain by a subduction zone. For the oceanic crust to be exposed as an ophiolite it must be lifted above sea level. This can happen in any one of several ways.

If the subducting plate is oceanic, it will take water down into the mantle in water-bearing metamorphic minerals such as zeolites and amphiboles. On being subducted these minerals will be heated and will release water, which will convert some of the peridotite in the overlying mantle into serpentinite. The serpentinization of the peridotite increases its volume and makes it lighter, so that it tends to rise. The process contributes to the uplift of the overlying crust and mantle (see Figure 8.6a).

Alternatively, if the potential ophiolite is underthrust by continental crust, the presence of lower-density continental rocks at depth will upset the normal equilibrium and cause uplift (see Figure 8.6b,c). In 1963 David Masson-Smith of the Institute of Geological Sciences and I, investigating variations in gravity over the Troodos massif, proposed that this ophiolite was lifted above sea level by the under-thrusting of continental crust. The process was intensified by the rise of a mass of serpentinite under what is now the center of the massif. Similar mechanisms involving the underthrusting of an eventual ophiolite by an oceanic plate are favored by my Open University colleagues working in Oman to explain that ophiolite and by Daniel E. Karig of Cornell University to explain the Zimbales ophiolite in the Philippines.

Support of the underthrusting mechanism comes from the study of the metamorphic rocks under ophiolites. Many ophiolites, notably those of Newfoundland and Oman, are underlain by a thin layer of metamorphosed rocks separating the ophiolite from the underthrusting material. These metamorphic rocks were formed in a zone where the temperature was highest (about 600 degrees C.) immediately adjacent to the overlying ophiolite. The temperature diminished rapidly with depth, so that only a few hundred meters below the top of the zone the rocks are unmetamorphosed. The metamorphic agent here is probably a combintion of heat emanating from the overlying slab of oceanic crust and frictional heat generated by the downward movement of the subducting plate. In these instances the assemblages of metamorphic minerals and their disposition are best explained by the continuous subduction of an oceanic plate.

In many of the less deformed ophiolites the rock layers that are correlated with the oceanic layers identified by earthquake-wave velocities are markedly thinner than the oceanic layers. This has led to the proposal that ophiolites represent oceanic crust that is thinner than normal, formed at minor ridges in small, marginal seas (see Figure 8.6*f*). Such thin crust, it has been argued, could be more easily obducted. Indeed, Julian Pearce and others have demonstrated that ophiolitic basalts show geochemical features most compatible with their being derived from a water-bearing melt whose water content had come from an underlying subduction zone. Moreover, the age of an ophiolite and the time of its emplacement are commonly very close together, and so it has been suggested that the oceanic crust formed first gets emplaced as an ophiolite before the conveyor belt of the subduction process gets well under way. This too could explain the thinness of an ophiolite layer.

Other factors, however, are not easily compatible with this simple model. There is no doubt that most ophiolites are in contact with underlying rocks of continental origin. In some instances it can be proved that it is the fragments of oceanic crust that have moved over the continental rocks. A classic case is the emplacement of the Papua-New Guinea ophiolite, which, as it has been described by Hugh L. Davies of the Australian Bureau of Mineral Resources, Geology and Geophysics has been emplaced southward along a thrust zone inclined to the north (see Figure 1.8). This zone, unlike all others around the Pacific, is inclined seaward; it is one of the few cases where the term obduction may be appropriate.

In other instances blocks of oceanic crust, often many kilometers across, are found in a mélange, embedded in a matrix of serpentinite or muddy sediment. In these instances it seems most likely the blocks of oceanic crust were detached as the oceanic plate buckled and fractured before it was subducted. The blocks detached by the process fell into an adjacent deep oceanic trench. Such trenches are the main surface indication of a subduction zone (see Figure 8.6*d*).

The term "obduction" is therefore a troublesome one. It is nonetheless useful as long as the complexities involved are kept in mind. Wht may be more to the point is that in the present cycle of plate tectonics (there have been other cycles in the past) ophiolites are found mostly at or near subduction zones. Hence their presence can be taken to indicate the proximity of fossil destructive margins. This association has been applied successfully in geological reconstructions of Mesozoic and Lower Paleozoic terrains by Robert Coleman, John Dewey and Alan Gilbert Smith, among others (see Figure 8.7).

"Ophiolite" too is a somewhat troublesome term for on-land fragments of oceanic crust. Certainly the term has changed its meaning. In geology before the emergence of plate tectonics an ophiolite was associated with the initial stages in the development of a geosyncline; it consisted largely of serpentinite and had been metamorphosed in the course of a cycle of mountain building. The Troodos massif shows none of these features, and in the 1950's, when it was first being studied in detail, it was not even considered an ophiolite. Then by the early 1970's it became apparent that there were enough structures like the Troodos massif for them to need a collective label. Many of the complexes that were already called ophiolites turned out on reexamination to have the same structure; hence it was perhaps inevitable that the label ophiolite was retained. The term was, however, given a new and far more precise meaning. Today most earth scientists accept that ophiolites are on-land fragments of oceanic crust formed at oceanic ridges or rises, and with this consensus the time seems ripe to accept the term ophiolite, however inappropriate it may seem.

The Structure of Mountain Ranges

What hold mountains up? Some stand on plates of strong rock; others are buoyed by crustal roots reaching deep into the mantle. The latter may collapse when their flanks are not pushed together.

• • •

Peter Molnar
July, 1986

In looking at mountains one is generally first struck by the topography: the extraordinary scale, the shapes carved by glaciers and streams, the contours smoothed and decorated by vegetation. Many people are inspired by the same awe they feel before certain manmade structures, such as the soaring arches and stained glass of a Gothic cathedral. Yet as the eye steps from detail to detail across a landscape it is easy to forget that enormous forces are required not only to build but also to support a mountain range. Each range, like each cathedral, stands on a foundation, without which it would collapse. If one wants to feel more than inarticulate wonder before mountains or buildings, it helps to understand the invisible mechanisms that support the visible beauty. Hence the purpose of this chapter: to describe the underlying structure — the tectonics, if not the architecture — of mountain ranges.

TWO KINDS OF SUPPORT

The analogy to architecture is not merely rhetorical; the different solutions architects have found to the problem of supporting buildings have parallels in the structure of mountain ranges. One solution is to build on a basement of strong, inflexible rock. Some of the world's tallest buildings, for example, stand on the Manhattan schist, a rock formation that has not been significantly heated or deformed (and thereby weakened) since the Precambrian era ended some 600 million years ago. The world's highest mountains, the Himalayas, are like the Manhattan skyscrapers: they stand on a thick shield of strong Precambrian rock, the northern edge of the Indian subcontinent.

A basement of strong rock, however, is not necessary for the support of a large structure. I work in a 20-story building in Cambridge, Mass., that rests on pilings pounded 40 meters into artificial fill and glacial moraine in what was once a tidal basin along the Charles River. To some extent the building floats on water-saturated deposits, and in that respect it is not unlike a large ship. Mountains too can be supported by the buoyancy of light material floating on heavier material. An example is the Tibetan plateau north of the Himalayas, nearly all of which lies above 4,500 meters. Unlike the substrata of the Himalayas themselves, the substrata of the plateau appear to be weak and easily deformed —

like the landfill under my office building or the water under a ship.

THE SURVEY OF INDIA

The Himalayas and the neighboring Tibetan plateau thus exemplify two quite distinct mechanisms for supporting mountain ranges (see Figure 9.1), which is not to say that both mechanisms cannot operate in the same range. In fact, it was the study of this area 140 years ago that led to the first advances in understanding the structure of mountains. The pioneers were the surveyor George Everest, J. H. Pratt, the scientifically inclined archdeacon of Calcutta, and George B. Airy, the eminent mathematical physicist and Astronomer Royal of Britain. The story of their work is in itself a fascinating bit of intellectual history.

In the 1840's Everest was directing the first topographical survey of the Indian subcontinent. His crews had two methods of measuring distances. First, they could measure short distances by the conventional surveying technique of triangulation, arriving at longer distances step by step. Second, they could determine the relative positions of two widely separated points directly by observing the position of a reference star from both points at the same time of day. In principle the two methods should have yielded similar results, but in practice there were large discrepancies. The most celebrated of these concerned the distance between the towns of Kaliana and Kalianpur, respectively some 100 and 700 kilometers south of the Himalayan front. The astronomical survey placed the two towns about 150 meters closer to each other than triangulation did.

Everest assumed that cumulative errors in triangulation accounted for the discrepancy, but in 1854 Pratt showed that the error lay instead with the astronomical measurements. To determine the position of a star on the celestial sphere the surveyors had to know precisely the direction of the zenith (the vertical direction), which was defined by a plumb line. Pratt suggested that the gravitational attraction exerted by the large mass of the Himalayas and the Tibetan plateau would deflect the plumb bob to the north, and the deflection would be greater at Kaliana than at Kalianpur because Kaliana is closer to the mountains. The resulting difference in the measured directions of the zenith would introduce an error into the calculation of the relative positions of the two towns.

When Pratt tried to determine the size of the error by estimating the mass of the Himalayas and the Tibetan plateau, he made a puzzling discovery. His results indicated that a plumb bob should be deflected by 28 seconds of arc at Kaliana and by 12 arc seconds at Kalianpur. The 16-arc-second difference may seem small, but actually it corresponded to an error in the astronomical distance measurement that was three times larger than the observed discrepancy of 150 meters. Apparently, Pratt concluded, the real difference in the gravitational deflection of the plumb bob was only about five arc seconds, which implied that he had overestimated the mass of the mountains; there was a lot less mass under the Himalayas and Tibet than his analysis of their topography had suggested. Indeed, had Pratt had access to accurate topographic maps (his maps put most of Tibet at just over half its true altitude) he would have inferred an even greater "missing" mass (see Figure 9.2).

CRUST AND MANTLE

Airy read Pratt's paper at his desk in London. At first the idea of missing mass surprised him, but then he quickly realized that the surface of the earth is probably not strong enough to support a huge mass of mountains without deforming in some way. The deformation leads to a mass deficit under the mountains that compensates the excess mass on the surface. Compensation of this type is familiar to most people as the phenomenon Archimedes discovered when he got into a full bathtub and it overflowed; to geologists it is now known as isostasy.

In Airy's conception of isostasy the earth's light crust floated on a heavier but weak, fluidlike substratum: the mantle. The chemistry of the crust is well known today, and it is indeed lighter than the

Figure 9.1 HIMALAYAS AND TIBET are shown in a photograph from the space shuttle *Challenger*. The view is toward the southeast, with Tibet in the lower left, the snowcapped Himalayas diagonally across the center and the green Ganges plain of India in the upper right. The area is about 175 kilometers wide. The large valley in the center is a graben formed where a block of crust has dropped down along normal faults as the adjacent crust has spread apart. The valley cuts through east–west-trending folds formed during the collision of India and Eurasia that raised the Himalayas. It continues to the south into the Kali Gandake valley, flanked by two peaks more than 8,000 meters high—Annapurna and Dhaulagiri.

mantle. Although both layers consist primarily of oxygen and silicon, the relatively heavy elements iron and magnesium are much more abundant in the mantle. In contrast, larger fractions of the relatively light elements, including sodium, calcium, aluminum and potassium, are concentrated in the crust. As a result the crust is less dense than the mantle, and so it is not unreasonable to think, as Airy did, that the crust floats on the mantle as cream floats on milk. Like the boundary between cream and milk, the boundary between crust and mantle is quite sharp.

Pratt shared Airy's conception of a floating crust, but the two men disagreed on the mechanism underlying isostatic compensation. Pratt thought the temperature and hence the density of the crust vary from place to place. Where the crust is hotter and lighter than average it rises to form mountains; where it is cold and dense it subsides to form vast lowlands. Airy, on the other hand, thought the density of the crust is farily uniform but that its thickness varies. The crust is thicker under mountains, he argued than under lowlands; the visible mountains are like the tips of icebergs, and like icebergs they are supported by deep, invisible roots.

Seismological studies over the past several decades have confirmed Airy's prediction that the thickness of the crust varies substantially. Continental crust is on the average between 30 and 40 kilometers thick, but under mountains the thickness may increase to as much as 75 kilometers. The crustal roots compensate the excess mass of the mountains by displacing denser mantle rock. Conversely, the crust under the deep oceans compensates the low density of water by being only about six kilometers thick. Gravity-induced rock movements keep the earth in approximate isostatic equilibrium, such that the mass of an imaginary column through the earth is roughly the same whether its surface is a mountain range or part of an ocean.

LITHOSPHERE AND ASTHENOSPHERE

In spite of such impressive confirmations Airy's version of isostasy is only approximately correct. Early in this century, even before seismologists had confirmed that the thickness of the crust varies, they had found that the mantle, like the crust, is solid rather than liquid. Hence the image of the crust floating on the mantle is an oversimplification, and the same must be true of Airy's hypothesis. In the 1930's the Dutch geophysicist Felix A. Vening-Meinesz suggested that the isostatic compensation of a topographic load should take place on a regional rather than a local scale and should involve more than just the formation of crustal roots. (In the limiting case this is obvious; the crust does not poke hundreds of meters into the mantle under the Empire State Building.)

Specifically, Vening-Meinesz proposed that a large load such as a mountain range deflects the earth's strong outer layer, now called the lithosphere. The lithosphere usually includes not only the crust but also the uppermost part of the mantle. It overlies a weak, fluidlike layer called the asthenosphere (see Figure 9.3). To a first approximation the lithosphere can be treated as elastic, and under a mountain range it bends downward, thereby distributing the weight of the range over a broad region. The bending of the lithosphere creates a trough parallel to the range. As a result the excess mass of the mountains is compensated in part by a mass deficit in the trough and not just by a deficit directly under the range.

The lithosphere, it is now known, is not a continuous layer but instead consists of 20 or so separate plates. The movements of the plates over the asthenosphere account for the formation of ocean basins and mountain ranges and for the other phenomena known collectively as plate tectonics. Although descriptions of such horizontal plate movements generally treat the plates as being rigid rather than elastic bodies, this does not contradict Vening-Meinesz' hypothesis that the lithosphere bends elastically under a topographic load. A lithospheric plate is somewhat like a wood table: the table moves rigidly when it is pushed across the floor, but it may sag in the middle when a heavy load is placed on it.

The asthenosphere offers buoyant resistance to the sagging of the lithosphere, but the lithosphere does not float on the asthenosphere. Unlike the boundary between crust and mantle, the boundary between lithosphere and asthenosphere is not a chemical one, and the density contrast across the boundary is therefore negligible. The buoyant restoring force exerted by the asthenosphere arises from its being much denser than the layer of air or water overlying the flexed lithosphere.

The lithosphere and asthenosphere differ in temperature rather than in composition: the lithosphere is cooler, which explains why it is stronger. Within the lithosphere temperature increases rapidly with depth, reaching a level of roughly 1,300 degrees at

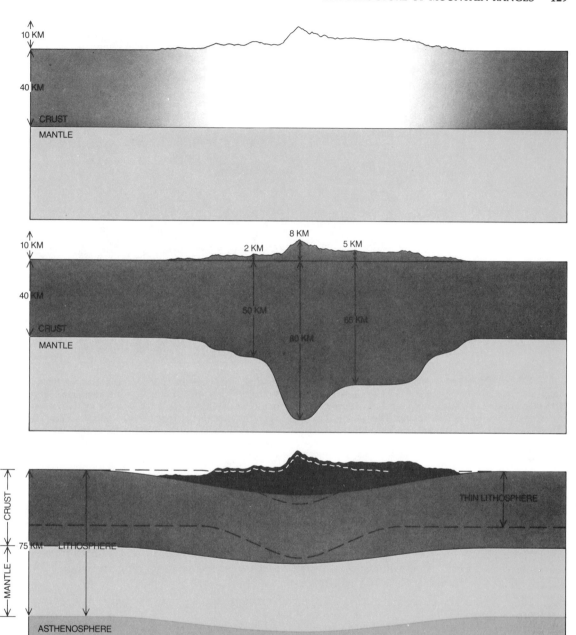

Figure 9.2 THREE MECHANISMS of isostatic compensation embody different ideas of how mountain ranges are supported. J. H. Pratt proposed (*top*) that the density of the crust varies laterally and that mountains stand high where the density is low (*light gray*). George B. Airy (*middle*) assumed that the crust is of uniform density but it is thicker under mountain ranges. Felix A. Vening-Meinesz proposed that a mountain range is compensated regionally rather than locally by lithosphere bending (*bottom*). Mountains stand higher on thick lithosphere because it bends less than thin lithosphere (*broken lines*).

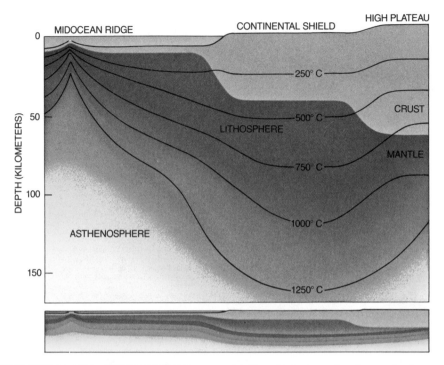

Figure 9.3 LITHOSPHERE AND ASTHENOSPHERE are shown in vertically exaggerated (*top*) and unexaggerated (*bottom*) cross sections. The lithosphere is cooler and therefore stronger than the asthenosphere. The thickness of the lithosphere varies considerably. Under midocean ridges, where it is created by the upwelling of hot material from the asthenosphere, it is very thin; under continental shields, which are made of crust that has not been heated for 600 million years or more, its thickness may increase to more than 150 kilometers. Under some high plateaus, however, the lithosphere is not thick. Indeed, under Tibet the thermal boundary between lithosphere and asthenosphere appears to fall within the crust instead of in the upper mantle.

the boundary of the asthenosphere. The boundary is not sharp like the chemical boundary between crust and mantle, and there is no general agreement among investigators on how to define it. What is clear is that like the crust the lithosphere varies widely in thickness, from as little as 10 kilometers to more than 150.

The thicker a wood table is, the greater the load it can support and the less it sags. The same is true of the lithosphere. A thick plate is stronger than a thin one and bends less under the weight of a mountain range. Consequently a range should stand higher on a thick plate, other things being equal, than it would on a thin one. High mountains can nonetheless exist on a thin plate if they are supported in the way Airy envisioned, by deep crustal roots. The isostatic mechanisms put forward by Airy and Vening-Meinesz are not mutually exclusive. On the contrary, it has been found that a mountain range can be supported by a strong foundation of thick lithosphere (like the New York skyscrapers), by deep roots of light crust (like a ship) or by a combination of both mechanisms. The relative importance of the mechanisms varies from range to range.

THE HIMALAYAS AND TIBET

To determine which mechanism is more important in the case of the Himalayas and the Tibetan plateau one must first consider how those mountains were formed (see Figure 9.4). Some 70 million years ago India and the rocks that now make up the Himalayas were about 8,000 kilometers south of their present position, drifting northward from Antarctica toward Asia on a large plate consisting primarily of oceanic lithosphere. Southern Tibet was at that time on the south coast of Asia and lay about 2,000 kilometers south of where it is now. As the

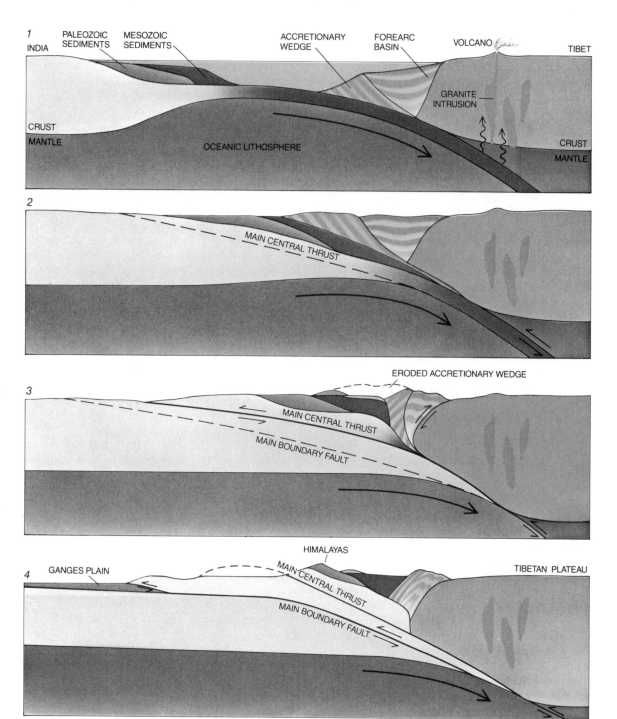

Figure 9.4 FORMATION OF HIMALAYAS is depicted schematically in simplified, exaggerated diagrams. The Indian plate was subducted beneath Tibet 70 to 60 million years ago, prior to collision with the Eurasian plate (1). Between 55 and 40 million years ago the two landmasses collided (2). The Indian continent presumably was too bouyant to subduct very far beneath Tibet and the Main Central Thrust formed as a result. The Main Central Thrust became inactive about 20 to 10 million years ago (3), and the Main Boundary Fault formed. Since then India has slid northward along the Main Boundary Fault (4), thrusting a second slice of crust onto the subcontinent and lifting up the Himalayas. Many of the peaks are capped by Paleozoic marine sediments.

Indian and Eurasian plates collided the oceanic lithosphere north of the Indian landmass was bent down and thrust under Tibet, much as the plates under the Pacific Ocean are now being thrust under Japan, the Aleutians and South America. (Geologists call this process subduction.) It was as if the Indian plate were a conveyor belt trundling around a spool under southern Tibet.

Sometime between 55 and 40 million years ago the Indian landmass itself struck the south coast of Asia, and at that point the conveyor belt began to jam; the speed of the Indian plate was reduced from between 10 and 20 centimeters per year to about five centimeters per year (the rate at which India still plows into Eurasia today). As India plunged under Tibet with tremendous force, a northward-dipping fault tore through the northern edge of the subcontinent. The crust under the fault plane continued to move northward and downward, but a slice of continental shelf and deep crust above the fault plane was in effect shaved off the oncoming subcontinent and thrust backward on top of it. Between 20 and 10 million years ago the process was repeated: the first fault became inactive and a second fault formed at a deeper level. A second slice of Indian crust was thrust onto the subcontinent, lifting up the first slice. The eroded remnants of these two slices of ancient Indian crust are exposed today in the Himalayas, and they constitute the bulk of the range.

The heavy weight of the Himalayas bends the Indian plate downward south of the range. Over millions of years sediments eroded from the mountains have filled the resulting trough, forming the broad plains of the Ganges and Indus rivers. Seismological and drilling results obtained by the Oil and Natural Gas Commission of India have documented the presence of the trough in the Precambrian bedrock under the sediments. The bedrock dips smoothly down toward the mountains, reaching a depth of about five kilometers at the front. Some 200 to 300 kilometers south of the front, at the edge of the trough, the bedrock is exposed at the surface.

Considering the great weight of the Himalayas the trough is not very deep. The Indian plate does not bend much because it is particularly thick and strong. Just how thick it or any plate is cannot be determined precisely, but by making some simplifying assumptions one can calculate the comparative thicknesses of different plates. Hélène Lyon-Caen of the University of Paris has shown, for example, that the Indian plate is more than twice as thick as the Pacific plate under Hawaii.

The strength and thickness of the Indian lithosphere are a major reason the Himalayan peaks are so high. They are definitely not supported by deep crustal roots in the manner proposed by Airy. The thickness of the crust under the Himalayas is only about 55 kilometers, which is more than the 35 to 40 kilometers observed under the rest of India but much less than the 80 or so kilometers that would be needed to support the mountains solely by crustal buoyancy. The Himalayas are a clear case in which Airy's notion of local isostatic compensation fails and Vening-Meinesz' notion of regional compensation through bending of the lithosphere is required.

The Tibetan plateau, on the other hand, does fit Airy's conception. The plateau extends north of the mountains for hundreds of kilometers, and only in a few valleys near the edges does its altitude drop below 4,500 meters. Seismological work by Wang-Ping Chen of the University of Illinois at Urbana-Champaign and Barbara Romanowicz of the University of Paris suggests the Tibetan crust is generally between about 65 and 70 kilometers thick —thicker than the crust under the Himalayan peaks. The weight of the high plateau is compensated primarily by the buoyancy of its deep crustal root, as Airy proposed 130 years ago.

OTHER RANGES

Other mountain ranges and high plateaus bear out the structural dichotomy exemplified by the Himalayas and Tibet. The abundant folding of rock layers in the Alps is evidence that they were formed in much the same way as the Himalayas, by crustal material that was sheared off the southern edge of Europe and overthrust northward onto the European plate when it collided with the Italian prong of the African plate. The Molasse basin in northwestern Switzerland is analogous to the Ganges and Indus plains: it is filled with debris eroded from the Alps, and its existence is due at least in part to the downward flexing of the European plate under the weight of the mountains. Garry D. Karner, now at the University of Durham, and Anthony B. Watts of the Lamont-Doherty Geological Observatory of Columbia University have shown that the European plate is less than half as thick as the Indian plate.

The difference probably helps to explain why the Himalayas are nearly twice as high as the Alps. The Himalayas stand on a stronger foundation.

The Rocky Mountains in Canada also rest on a downward-flexed lithospheric plate. The exact manner in which the Rockies formed is still a matter of debate. It is clear, however, that the Canadian Rockies consist of slices of sedimentary rock that were successively detached from the underlying basement rock and thrust eastward on top of one another. A detailed exploration of this area has shown that the basement rock, part of the Precambrian Canadian shield, dips gently toward the west under the mountains. The dip in the lithosphere indicates that the weight of the mountains is compensated regionally, as Vening-Meinesz predicted.

Although they were formed in an altogether different way, the Hawaiian Islands provide another example of regional compensation (see Figure 9.5). The islands were built up volcanically, by molten rock rising from the asthenosphere through the Pacific lithosphere and pouring out onto the ocean floor. The resulting peaks are gigantic: Mauna Kea on Hawaii stands more than 4,200 meters above sea level and some 9,000 meters above the surrounding ocean floor. The weight of the islands bends the Pacific plate downward by a few hundred meters, creating a "moat" around the islands (see Figure 2.6). Just outside the moat the plate is warped slightly upward. The upwarping arises because the asthenosphere resists the bending of the lithosphere.

Whereas the Alps, the Canadian Rockies and the Hawaiian Islands have foundations similar to those of the Himalayas, the Andes, the highest mountains in the Western Hemisphere (see Figure 9.6) are more akin to Tibet. The weight of the range seems to be supported by a buoyant crustal root as much as 70 kilometers deep. Indeed, the Andean crust has been the focus of debate on one of the main unresolved questions about mountain building: the question of how (as opposed to why, which Airy explained) the crust is thickened under many ranges.

There are two conceivable answers. First, the crust can be thickened by volcanic magma that wells up from the mantle and cools in the crust, forming intrusions of granite and other igneous rocks. Second, a block of crust becomes thicker if its edges are pushed together by horizontal forces and it is thereby shortened. In the Andes intrusive volcanism and crustal shortening occur side by side; the question is which process contributes more to the thickening of the crust.

The western cordillera of the Andes is a volcanic arc of the kind typically found above a subduction zone, where one lithospheric plate dives under another. As the Pacific-crust-bearing Nazca plate dives into the asthenosphere it is heated, and molten rock—either from the plate itself or from the asthenosphere above it—rises into the crust of the overriding South American plate, forming volcanoes and granitic intrusions. Consequently rocks in the western Andes and on the coastal plains of Peru and Chile are predominantly volcanic. Most of the rocks in the high central plateau and in the eastern cordillera, however, are not volcanic. Instead they are primarily sedimentary rocks folded and thrust on top of one another (see Figure 9.7). The folding and overthrusting is evidence that the crust in those regions has been shortened in a direction perpendicular to the range.

Crustal shortening continues today on the eastern flank of the Andes. Seismograms analyzed by Douglas S. Chinn and Bryan Isacks of Cornell University, Gerardo Suárez of the National Autonomous University of Mexico and William Stauder of St. Louis University indicate that earthquakes on the eastern flank occur along faults where the Brazilian continental shield is thrust westward under the mountains. The rate of underthrusting is apparently only a few millimeters per year, but it may have been higher in the past. I have argued, along with Suárez, Lyon-Caen and B. Clark Burchfiel of the Massachusetts Institute of Technology, that this crustal shortening, and not volcanism, is largely responsible for the thick crust under the eastern cordillera. As the Andes are squeezed by the eastward thrust of the Nazca plate and the westward thrust of the Brazilian shield, the crust is thickened.

THE MOUNTAINS ARE FALLING

The buoyancy of the crustal root supports the weight of the mountains, but the horizontal forces that created the root also seem to have a more direct effect. It appears that they help to buttress the Andes and prevent the range from spreading and collapsing. Ironically, evidence for this view comes in part from the observation that the buttresses are beginning to fail. While the crust on the sides of the range is being pushed together, the crust in some

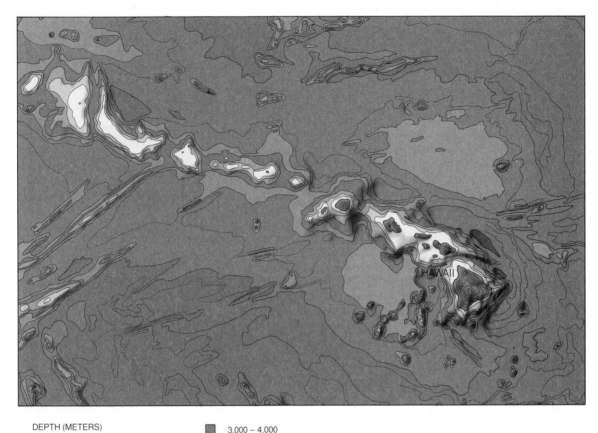

DEPTH (METERS)		
ABOVE SEA LEVEL		3,000 – 4,000
0 – 1,000		4,000 – 4,500
1,000 – 2,000		4,500 – 5,000
2,000 – 3,000		5,000 – 6,000
		BELOW 6,000

Figure 9.5 BENDING OF THE LITHOSPHERE is evident in this bathymetric chart of the region around the Hawaiian Islands. The islands are volcanic structures built of lava erupted onto the sea floor. Their weight bends the lithosphere downward by several hundred meters, form- ing a "moat" that is deepest around the island of Hawaii. A broad region of the sea floor around the islands is probably swelled by an upwelling of hot rock in the asthenosphere; to the north and south of the Hawaiian region the ocean gets progressively deeper.

regions of the high Andes is being pulled apart. The Cordillera Blanca, a western chain that includes Peru's highest peak, Huascarán, is a good example (see Figure 9.8). The chain is bounded on its western side by a steep fault running parallel to the range; to the west and the fult the crust has dropped and moved away from the mountains. This type of fault, along which one block of crust drops in relation to another, is called a normal fault. It is a clear indication that the crust is extending.

My introductory analogy between a mountain range and a Gothic cathedral may help to clarify the significance of the normal faults in the Andes. The Andean peaks and high plateaus, the roof of the Western Hemisphere, are like the vaulted ceiling of a cathedral. The vaults exert outward thrusts on the walls that tend to push the walls apart. (In the case of the cathedral the thrust is due not only to gravity but also to wind loading.) To prevent the ceiling from collapsing, Gothic architects built huge flying buttresses that countered the outward forces on the walls. Another way of solving the problem,

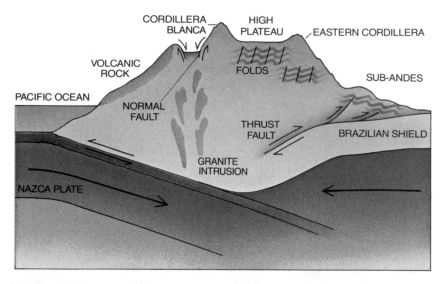

Figure 9.6 PERUVIAN ANDES are supported by a deep crustal root. Under the western Cordillera Blanca the crust has been thickened by intrusions of volcanic material rising above the Nazca plate as it plunges under South America. The convergence of the two plates also thickens the crust by pushing it together, or shortening it. Folded rock formations in the eastern sub-Andes prove that the crust there is being shortened and lifted up as the Brazilian shield is thrust under the mountains. Although the sides of the range are still being pushed together, the crust in the high Andes is stretching: on the western side of the Cordillera Blanca great blocks of crust have dropped down along normal faults.

which a present-day architect might favor, would be simply to stretch steel cables between the opposing walls; the cables would have enough tensile strength to hold the walls together.

In a sense the Nazca plate and the Brazilian shield are the flying buttresses of the Andes. Their inward horizontal thrust on the flanks of the range helps to support the high peaks and plateaus. What the normal faults in the high Andes suggest is that the horizontal buttressing forces are no longer strong enough to discharge their function satisfactorily; nor are the rocks that make up the mountains strong enough to serve the function of steel cables and hold the mountains together. Although the crust is still being pushed together on the eastern flank of the mountains, in the high Andes it is being stretched apart: the roof is falling. The range as a whole may be entering a stage of decline that will ultimately lead it to collapse completely under its own weight.

If the Andes do collapse, they will probably not have been the first mountains to suffer such a fate. Many workers, myself included, consider the Andes a modern analogue of a mountain range that domi-nated the western U.S. between 80 and 30 million years ago, when the Rockies were just being built farther east. At that time a lithospheric plate under the eastern Pacific Ocean was converging with the North American plate, and the oceanic lithosphere was being subducted, just as the Nazca plate is now being subducted under South America. Sometime between 30 and 10 million years ago the subduction under North America stopped. With the cessation of convergence the horizontal forces that had shortened and thickened the crust under the mountain range would have diminished or even disappeared.

When that happened, the crust would have begun to spread. Signs of crustal extension are plentiful in the Basin and Range province west of the Rockies, between central Utah and the Sierra Nevada: the alternating basins and tilted ranges are bounded by normal faults like the ones west of the Cordillera Blanca. As the crust has spread apart, blocks of crust have dropped along the normal faults to form the basins. (One of the basins is Death Valley, which is now below sea level but may once have been at an altitude of several kilometers.) According to this view, the Basin and Range province

Figure 9.7 TILTED AND FOLDED sedimentary rock in the high plateau of the Andes indicates the plateau was formed by shortening of the crust. The view is to the southeast from east of the Cordillera Blanca. The folded limestone, sandstone and shale layers were pushed together from the east and west. The rocks in the distant snowcapped mountains, part of the western cordillera, are also folded.

is the remnant of a broad belt of mountains and high plateaus that collapsed after the horizontal forces supporting the belt were removed. One day the Andes may look like the Basin and Range.

Tibet too may be collapsing. Although the pressure applied by India's northward motion to the rest of Asia seems to be enough to prevent Tibet from extending in a north-south direction, the high plateau has no similar buttress on its eastern flank. Accordingly the plateau is laced with northerly trending normal faults, along which one side moves down and away from the other. Tibet is spreading to the east, and in the process it is pushing southeastern China eastward with respect to the rest of Asia.

Why are the Andes and Tibet in particular susceptible to collapse? It is precisely because they are supported mainly by deep crustal roots. The strength of crustal rock decreases rapidly with increasing temperature and hence with increasing temperature and hence with increasing depth, probably more rapidly than the strength of mantle rock. Thick crust therefore tends to be weak. Moreover, for reasons that are not fully understood, the crust and upper mantle under Tibet and the Andes seem to be comparatively warm; indeed, the boundary between the cool lithosphere and the warm asthenosphere may actually lie within the crust rather than well below the crust, as it does in most areas. As a result the crustal roots of Tibet and the Andes appear to be weak and fluidlike, and to the extent that they are not restrained by horizontal buttressing forces they tend to spread out. The horizontal forces are what keep the roots deep and the plateaus high. Paul Tapponnier of the University of Paris and I have suggested that the plateaus can be thought of as pressure gauges: the more they are subjected to horizontal pressure, the greater their elevation is. In Tibet and the Andes the pressure may have begun to drop.

Figure 9.8 NORMAL FAULTS are responsible for the steep western face of the Cordillera Blanca. The peak on the left, Huandoy, is 6,356 meters high. The escarpment running along the base of the mountains and across mo-raines left by a receded glacier is an active normal fault. In relation to the mountains the valley in the foreground has dropped several kilometers along this and parallel faults. Apparently the range is collapsing as the crust spreads.

The Himalayas, the Alps and the Rockies, on the other hand, are supported primarily by strong, thick lithosphere consisting of relatively cold crust and mantle. (The crust under the Himalayas, for example, is much colder than the Tibetan crust because it is underthrust by the cold Indian plate.) Although these ranges were formed by horizontal forces, they have no need of horizontal buttressing to remain standing. Judging from the general absence of nor-mal faults in them, they do not seem to be collaps-ing.

DYNAMICS

Some mountain ranges are like pressure gauges, others are like loads on elastic plates — the analo-gies are correct as far as they go, but I must stress that they are simplifications. At a finer level of detail they reveal their limitations. Karner and Watts have shown, for instance, that the weight of the Alps is not great enough to bend the European plate down as much as it is under the Molasse Basin; some additional force must pull the plate down. In contrast, Lyon-Caen and I have found that the strength of the Indian plate, great as it is, does not fully account for the high altitudes in the Himalayas; an additional force apparently flexes the northern edge of the plate up. Hawaii also is pushed up. The depth of the ocean floor in a large area around the islands outside the Hawaiian moat is only about 4,500 meters, whereas 1,000 kilometers or so to the north and south the ocean depth is about 5,500 meters.

These deviations indicate that the simple model of a plate bending under the weight of the mountains is incomplete. What is missing is a consideration of plate dynamics, of the forces that drive continents together, shorten the crust and cause huge terrains to be thrust onto the edges of strong plates. Plate motions are widely thought to be the surface manifestations of a convective circulation that extends deep into the mantle, but the overall pattern of the circulation is not well known.

Nevertheless, some conclusions can be drawn. It seems clear, for example, that Hawaii lies above a region of the asthenosphere where hot material wells upward. Some of the material erupts at the volcanoes on the islands, but the upwelling column is much broader than the islands themselves. The upward thrust of the hot material accounts for the broad swell in the sea floor around Hawaii.

Under other mountain ranges one might expect to find a downwelling of relatively cold material. Under the Himalayas the Indian plate, stripped of the crustal slices that make up the mountains, may be plunging into the asthenosphere. The material in the upper part of the plate is significantly colder and therefore denser than the asthenosphere, and so it should sink. The weight of the sinking material may help to pull the plate down. At the same time, as Lyon-Caen and I have contended, the part of the platte just behind the leading edge would be flexed upward, which would help to push the mountains up. (To visualize the phenomenon take a plastic or metal ruler and bend one end over the edge of a table.) The sinking material might also drive a circulation in the mantle that helps to push the Indian and Eurasian plates together.

MEASURING GRAVITY

How can one study the dynamics of the mantle and determine in particular if dense sinking material is present under mountain ranges? One method is to measure variations in the earth's gravity field; the field should be slightly stronger above regions of the earth that are underlain by dense material. Unfortunately the differences in gravity caused by density variations in the mantle are small, probably less than about .01 percent of the average value of 9.8 meters per second squared. In a mountainous region they are masked by the much larger differences caused by variations in topography. To correct for topographic effects one must have extremely accurate maps, which in areas such as the Himalayas are simply not available.

The solution is probably to measure gravity with satellites. A satellite travels far above the gravitational influence of ridges and valleys, but its orbit is slightly perturbed by gravitational anomalies resulting from density variations in the mantle. By tracking the perturbations one can map the gravity field and the density variations. So far only large-scale gravity anomalies, thousands of kilometers wide and unrelated to mountain ranges, have been mapped. With improved tracking or, alternatively, with new satellite-borne instruments that directly measure lateral variations in gravity, it should eventually be possible to detect the smaller anomalies caused by density variations under mountain ranges.

When such measurements become available, a major step will have been taken toward understanding mountains not as static features but as features that grow and decay, as elements of an evolving earth. A deeper appreciation of the dynamics of mountain ranges will undoubtedly force a modification of some of the simple concepts I have presented here. Until then geophysicists fascinated by mountain architecture will remain in a situation not unlike that of the Gothic architects, who found they could support giant cathedrals with flying buttresses but who never really understood the underlying physical principles.

The Southern Appalachians and the Growth of Continents

A large seismic-reflection survey suggests that for at least half of the history of the earth continents have evolved by the stacking and shuffling of relatively thin sheets of material at their margins

• • •

Frederick A. Cook, Larry D. Brown and Jack E. Oliver
October, 1980

The theory of plate tectonics, which regards the crust of the earth as the upper part of a set of interacting rigid plates, has done much to describe the evolution of the ocean floor but has left much unexplained about the formation and structure of the continents. This state of affairs is beginning to change with the intensive application of the petroleum industry's exploration technique of seismic-reflection profiling to the study of the deep crust and the underlying upper mantle. The technique, which depends on the reflection of sound waves by discontinuities in the density of the rock and in the velocity of sound through it, has mapped new details of the geological structure of the continental basement.

The application of the technique in the southern Appalachian area of the U.S. reveals how the margins of continents change as ocean basins close and continents collide at subduction zones: the deep oceanic trenches near continental margins where oceanic crust plunges into the mantle. It has long been suspected that the continents grow by the accretion of crustal material at the subduction zones.

The seismic-reflection profiles of the southern Appalachians and related geological studies show how this actually happens: continents evolve out of the stacking and shuffling of thin horizontal sheets of crustal material.

The seismic profiles were made by the Consortium for Continental Reflection Profiling (COCORP), a group of geologists and geophysicists from universities, industry and government that is headed by Sidney Kaufman of Cornell University and two of us (Brown and Oliver (see Figure 10.1). COCORP collected its first data in Hardeman County, Tex., in 1975. Since then it has made profiles at 11 sites. By probing the crystalline rocks below sedimentary strata to depths of as much as 50 kilometers, the study has focused on the continental basement. The effort has been fruitful. COCORP has mapped a mid-crustal body of magma (liquid rock) in central New Mexico, explored an ancient buried rift valley in Michigan and traced a major thrust fault deep into the crust in southwestern Wyoming. Before the survey in Wyoming was done the depth and attitude of the fault, which bounds an important feature of the Rockies, had been the subject of much controversy.

The most spectacular finding was made in the southern Appalachians. The profiles revealed that the mountains are underlain to a depth of at least 18 kilometers by horizontal layers of material that is sedimentary or once was. Sedimentary rocks, of course, are one of the three broad types of rock, the other two being igneous and metamorphic. Sedimentary rocks originate with sediments deposited at the surface by water and wind. Igneous rocks are formed by the congealing of magma. Metamorphic rocks are formed when sedimentary of igneous rocks are subjected to heat and pressure over long periods of time. Metamorphic rocks can be described as high-grade or low-grade, depending on the degree to which they have been recrystallized in the process of metamorphism.

Most of the rocks at the surface in the southern Appalachians are highly deformed metamorphic ones. Furthermore, they are older than or contemporaneous with the horizontal sedimentary strata that were discovered under them. This fact suggests that roughly 475 million years ago the surface rocks began to be transported as a thin sheet for at least 260 kilometers over the eastern continental margin of the land mass that was to become North America. The discovery confirms the hypothesis that horizontal or near-horizontal thrusting can carry large volumes of crustal material great distances, and it suggests that a continent could grow and evolve by the emplacement of thin horizontal slices of material at the continental margins.

The seismic-reflection profiles may also have a more immediate kind of import. They have led geologists to speculate that there may be undiscovered deposits of oil or gas in the sedimentary rocks under part of the overthrust sheets. Sedimentary rocks are notably the best target for oil and gas exploration, because unlike igneous rocks and most metamorphic ones they have not been subjected to the pressures and temperatures that would destroy or expel the hydrocarbons.

T he first seismic-reflection studies of sedimentary basins were made in the 1920's. Since then the technique has been developed intensively by the petroleum industry as part of its effort to locate economic oil- and gas-bearing formations. The theory of the technique is quite simple. Sound waves generated by an explosion or some other source of acoustic energy at the surface radiate downward to impinge on discontinuities in subsurface layers of rock. The waves are reflected wherever there is an abrupt change in the rock's density or in the waves' velocity. The reflected energy is detected at the surface by an array of geophones, or vibration sensors.

The time it takes the wave to travel from the source to the reflecting discontinuity and then back to the surface depends on the depth of the discontinuity. Additional information about the discontinuity can be deduced from the recordings gathered by an array of geophones that respond to waves whose paths are almost vertical. The seismic-reflection profiles that are used to locate oil and gas respond to waves whose travel times are less than four or five seconds, corresponding to depths of between eight and 10 kilometers. Waves with longer travel times are mainly bouncing off rocks that are too deep to be reached by current techniques for recovering oil. The COCORP survey worked with waves whose travel times were as much as 20 seconds; such times correspond to depths of 60 or 70 kilometers.

COCORP did a multichannel reflection survey in which as many as 2,304 geophones (arranged in 96 groups of 24) were laid out over a distance of more than 6.5 kilometers. The entire spread of detectors was connected by cable to a truck in which the collected data were digitally stored on magnetic tape for the subsequent processing by computers that would construct a seismic image of the subsurface geology. If the wave sources and the geophones are positioned properly and the data are processed correctly, undesirable noise will be suppressed and the reflected signals will be enhanced.

Although explosives still sometimes serve as the source of acoustic energy in explorations for oil and gas, COCORP and much of the petroleum industry have turned to a nonexplosive, environmentally acceptable source named Vibroseis, which was invented at the Continental Oil Company in the 1950's. In a typical COCORP survey four or five truck-

Figure 10.1 SATELLITE IMAGE OF THE APPALACHIANS to the west of Roanoke, Va., is a false-color one made in the fall. Roanoke is the light green area to the right below the middle. The red areas are vegetation, the bright bluish green areas rivers and lakes. Here the mountains, which reach heights of between 500 and 1,000 meters, shift from a northeasterly trend to a more northerly one. The linear trends constitute the sedimentary material of the Valley and Ridge province. Seismic-reflection survey by COCORP covered part of this area and terrains of crystalline rocks to the southeast.

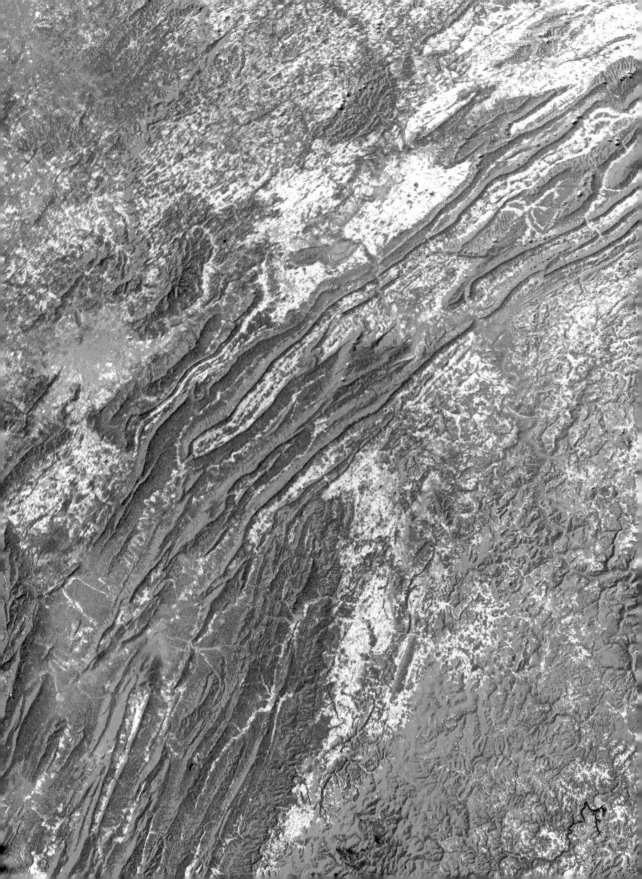

mounted vibrators direct into the ground a signal of between eight and 32 hertz. Over the sweep period of 30 seconds the frequency is made to vary linearly to form a "chirp." After each sweep the vibrators are moved forward a few meters and another sweep is made. After 16 sweeps covering a distance of 122 meters the process is repeated at the next source station 134 meters away.

The purpose of all of this is to record many reflections from the same sub-surface point (see Figure 10.2). The redundancy of the data enables the computer analysis to suppress noise, to enhance the strength of the reflected signal and to estimate the velocity of the subsurface vibrational waves. A seismic image of the structure of the continental basement is then constructed from the readings taken from many different configurations of sources and sensors. The process is somewhat like the medical technique of computed axial tomography (the "CAT scan"), in which X-ray readings taken from many different angles are combined to yield a representation of the internal structure of the living body in cross section.

In seismic-reflection profiling it often takes a few months to gather and process the data for a particular site. The sources and detectors are moved across the surface at a rate of between one and four kilometers a day. To reconstruct the subsurface geological structure requires the processing of enormous quantities of data. For example, the analysis of a major northwest-to-southeast traverse in Georgia and Tennessee called for the manipulation of roughly two billion items of information: 3,843 vibration points multiplied by 96 channels multiplied by a recording time of 50 seconds divided by a sampling rate of .008 second.

The computer analysis of the COCORP data included the demultiplexing of the multichannel field recordings, the elimination of particularly noisy data, the calculation of wave velocities, the collection of all reflection signals from each common depth point, the compensation for differences in the distance between sources and receivers, the adjustment for differences in topography and near-surface geology and the stacking, or superposition, of the several coherent signals for each common reflection point. The result of this intensive processing is a seismic cross section that resembles a geological cross section except for the important differences that depth is represented not by distance but by travel time and that lateral variations in the velocity

of sound in the rocks can distort the geometry of the reflections.

To infer an accurate geological cross section from the seismic section requires substantial skill and experience. For example, dipping interfaces are represented on the seismic cross section by reflections that seem out of position; the steeper the dip, the greater the "misplacement." Reflections often must be "migrated" back to give an accurate picture of the actual subsurface geology. (The COCORP profile of the southern Appalachians did not show many steep dips, so that migration was not much needed.)

Such was the exploration technique that was applied to the southern Appalachians. The Appalachians extend more than 3,000 kilometers from Newfoundland to central Alabama (see Figure 10.3). The southern Appalachians consist of a series of distinct geological provinces and belts trending from northeast to southwest. (A province is a general term for any region whose rocks have a similar history; a belt is a long, linear feature whose rocks have a similar composition.) From northwest to southeast the southern Appalachians are made up of the Valley and Ridge province, the Blue Ridge province, the Piedmont province (including the Inner Piedmont, the Charlotte belt and the Carolina slate belt) and the coastal plain (see Figure 10.4).

The Valley and Ridge province is characterized by folded and thrust-faulted strata of mostly unmetamorphosed sedimentary rocks formed between 600 million and 300 million years ago. The thrust faults and folds indicate that the rocks were much compressed in the horizontal direction. For a long time it was not known whether the deformation in the Valley and Ridge province involved the crystalline rocks forming the basement under the sedimentary strata (in which case the process would be called

Figure 10.2 NINE-STEP PROCESSING OF DATA in Vibroseis experiments typical of COCORP surveys is illustrated schematically. The thick black and red lines in the top three diagrams show the path of a signal reflected by a point P for three different positions of truck-mounted vibrators. Data processing involves nine steps (bottom). Recorded signals are displayed (1) and noisy signals are edited out (2). Reflections from a common depth point (e.g., P in upper diagrams) are collected together (3). Strong signals are muted (4) and common signals are lined up (5); common signals are collected in one trace (6). The single trace is shrunk to a sharper peak (7) and the shrunken summed signals are displayed (8). The data are then interpreted (9).

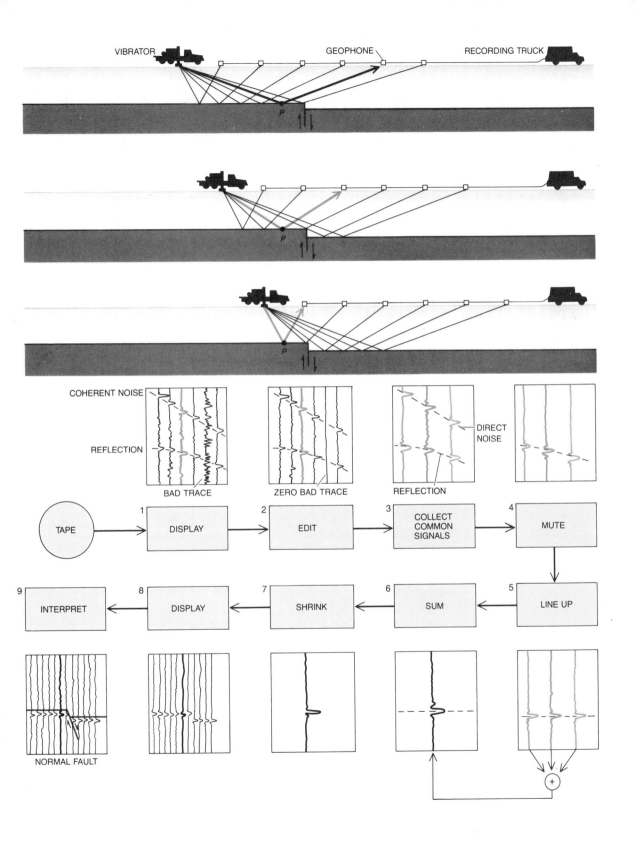

VIBRATOR   GEOPHONE   RECORDING TRUCK

_P_

_P_

_P_

COHERENT NOISE

REFLECTION

BAD TRACE   ZERO BAD TRACE   REFLECTION   DIRECT NOISE

| TAPE | 1 DISPLAY | 2 EDIT | 3 COLLECT COMMON SIGNALS | 4 MUTE |

| 9 INTERPRET | 8 DISPLAY | 7 SHRINK | 6 SUM | 5 LINE UP |

NORMAL FAULT

+

thick-skinned tectonics) or whether the deformation was confined chiefly to the sedimentary strata over the basement (thin-skinned tectonics). Now it is known from seismic-reflection profiles and explorations for oil and gas that the deformation is predominantly thin-skinned. Like a wrinkled rug on a floor the sedimentary strata seem to have ridden westward on top of large, horizontal detachment zones over the crystalline basement.

Southeast of the Valley and Ridge province is the Blue Ridge province. The main transition between the two is a large thrust fault that dips to the southeast. Unlike the sedimentary rocks of the Valley and Ridge province, the rocks of the Blue Ridge province have generally undergone much metamorphism. In both provinces however, the rocks have been deformed in three orogenies, or episodes of mountain building: the Taconic, the Acadian and the Alleghenian.

The basement of the Blue Ridge province includes Precambrian rocks that are at least a billion years old. Many geologists cited this fact as evidence that the Blue Ridge province is rooted in place, making it part of the main body of basement rock that forms the crust there. New data show, however, that although unmetamorphosed sedimentary rocks are rare in the province, they are occasionally found in "windows": areas where the metamorphic rocks have eroded away to expose the underlying material. The windows are mostly in the western Blue Ridge, and so it is now known that at least there the crystalline basement rocks are underlain by layers of unmetamorphosed sedimentary rocks. Moreover, one of the largest windows, the Grandfather Mountain window of North Carolina, is in the eastern Blue Ridge. This area, which exposes Cambrian carbonate rocks and shales, demonstrates that sedimentary rocks that have undergone little metamorphism are present under the eastern Blue Ridge as well as under the Western part.

At the eastern edge of the Blue Ridge province is a major topographic feature that extends from Alabama to Virginia. This feature, called the Brevard zone, is a narrow belt of multiply deformed rocks that marks the boundary between the Blue Ridge and the Piedmont. The history and the nature of the deformation in the Brevard zone have been the subject of much debate. In fact, COCORP did its first survey of the area in the hope of getting new data so that the debate could be resolved. One of the most important geological findings in the Brevard zone

was made by Robert D. Hatcher, Jr., of the University of South Carolina. He interpreted sedimentary rocks of an unusually low metamorphic grade as evidence of sedimentary strata deep under the Brevard Zone, with the rocks having been brought to the surface by faulting. This was additional evidence that sedimentary material may underlie the Blue Ridge province at least as far east as the Brevard zone.

Southeast of the Brevard zone is the Inner Piedmont, which consists primarily of high-grade metamorphic rocks intruded by bodies of igneous rock such as the Stone Mountain granite and the Elberton granite. The Piedmont has traditionally been thought to be the metamorphic "core" of the southern Appalachians. Because much of the metamorphic material seems to be sedimentary rocks that were extensively deformed, geologists believed the Piedmont had been vertically uplifted and the deformation had propagated westward into the Blue Ridge province and the Valley and Ridge province. The new seismic data have completely changed this view.

The Inner Piedmont is flanked on the southeast by the Kings Mountain belt, a narrow band of metamorphosed sedimentary rocks and volcanic rocks, some of which may be the remnants of a closed ocean or a geological basin at the continental margin. The major folds of the belt and the folds of the area to the southeast are aligned in a northeasterly trend, similar to the folds of the Inner Piedmont and of the Blue Ridge. The belt was scarred by at least two periods of deformation, one probably about 450 million years ago and the other about 350 million years ago.

The southeastern Piedmont is made up of the Charlotte belt, which consists of metamorphosed sedimentary material, and the Carolina slate belt, which consists primarily of metamorphosed volcanic material. The Carolina slate belt seems to be the remnant of a volcanic arc like the ones that are active today in the western Pacific. Volcanism in the belt probably started in the late Precambrian between 700 million and 650 million years ago and continued into the Cambrian until about 500 million years ago.

The coastal plain southeast of the Piedmont consists of a sequence of young sediments (less than 200 million years old) overlying a crystalline basement. Wells penetrating the basement have unearthed metamorphosed sedimentary rocks like

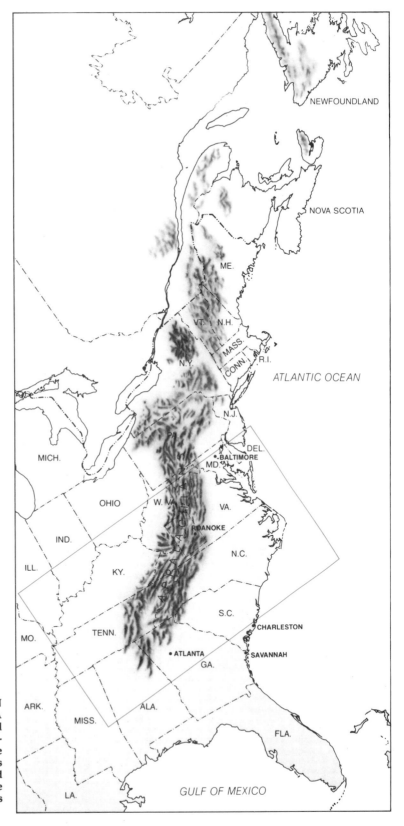

Figure 10.3 MAP OF THE EASTERN U.S. AND SOUTHERN CANADA shows how the Appalachians extend from central Alabama to Newfoundland. The broken rectangle on the map marks the Roanoke area that is shown in Figure 10.1. The colored rectangle outlines the area of the southern Appalachians, which is shown in Figure 10.4.

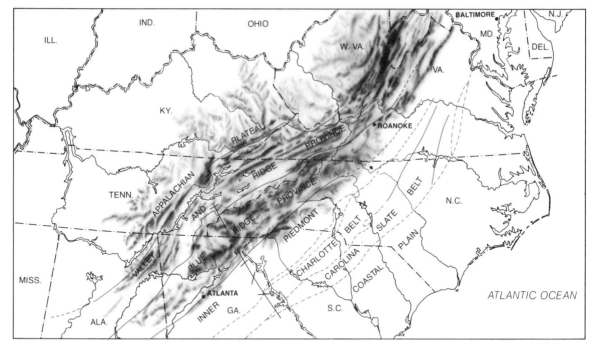

Figure 10.4 SOUTHERN APPALACHIANS are made up of four major topographic features trending from northeast to southwest: the Valley and Ridge province, the Blue Ridge province, the Piedmont province (including the Inner Piedmont, the Charlotte belt and the Carolina slate belt) and the coastal plain. COCORP studied the continental basement of the area along black line. Red lines mark thrust faults.

those of the Inner Piedmont and metamorphosed volcanic rocks like those of the Carolina slate belt. Perhaps the metamorphosed sedimentary and volcanic basement of the Piedmont and the Carolina slate belt extends below the coastal plain.

According to the historical picture developed by the theory of plate tectonics, the continents that now border the Atlantic were joined 200 million years ago like the pieces of a jigsaw puzzle to form one huge expanse of land. At that time North America began to separate from Europe, Africa and South Ameria. Since the continents were once connected, geological data from the eastern side of the North Atlantic may help in the interpretation of the structure and formation of the Appalachians. Although most plate-tectonic models of the Carboniferous period before the Atlantic started to open put western Africa adjacent to southern North America, a recent interpretation of paleomagnetic data by Edward Irving of the Department of Energy, Mines and Resources of Canada suggests that it was

northern South America that was then adjacent to eastern North America. In any event western Africa and northern South America both have belts of folding and thrusting that were probably created in the same Carboniferous orogeny.

The Mauritanide mountain chain of western Africa is characterized from east to west by a series of belts that are similar in some ways to the Appalachian belts. The eastern Mauritanides are made up of unmetamorphosed sedimentary strata partially covered by metamorphic rocks that have overridden the sediments from the west along thrust faults. To the west are older high-grade metamorphic rocks that resemble those of the southern Appalachian Piedmont. A coastal plain of horizontal younger rocks covers the rest of the orogen. Geological mapping of the area suggests that a mild episode of deformation about 550 million years ago was followed by a period of metamorphism and thrusting before the opening of the Atlantic. This period probably corresponds to the Alleghenian orogeny in the Appalachians. In a broad sense the Mauritanides of western Africa are a mirror image of the Appalachians.

Cocorp investigated the Appalachians in Tennessee, North Carolina and Georgia. The initial survey covered an area from about 100 kilometers southwest of Knoxville, Tenn., to about 100 kilometers northwest of Augusta, Ga. The most prominent finding is a southeast-dipping layer of reflections extending from the Valley and Ridge province under the Blue Ridge and the Inner Piedmont between four and 10 kilometers below the surface (see Figure 10.5).

Although several types and configurations of rock could give rise to the reflections, we interpret the reflecting materials as layered sedimentary strata. The interpretation is based on four items of evidence. First, the reflections under the Blue Ridge and the Inner Piedmont can be traced to and correlated with similar units of the Valley and Ridge province in which the rocks are known to be sedimentary. In fact, parts of the seismic-reflection profile of the Blue Ridge and the Piedmont are quite similar to the seismic-reflection profile of the Valley and Ridge. Second, the presence of sedimentary rocks in the windows of the Blue Ridge indicates that the crystalline rocks there overlie sedimentary material. Third, the discovery of unusual carbonate rocks in the Brevard zone suggests that they were scraped from underlying sedimentary layers by activity along the edges of the fault. Last, the cocorp results resemble those from seismic-reflection surveys of the current continental margins, which of course consist of sedimentary material.

H. Clark and his fellow workers at the Virginia Polytechnic Institute have made limited seismic-reflection studies of the area 100 kilometers north of the cocorp-survey area in North Carolina. The strata are also present under that area of the Blue Ridge. Therefore the strata probably underlie much of the southern Appalachians. Leonard Harris and Ken Bayer of the U.S. Geological Survey have analyzed recently collected seismic-reflection data from North Carolina that again show layered strata extending from the Valley and Ridge province under the Blue Ridge into the Piedmont.

The only explanation for the buried strata is that the overlying crystalline rocks were emplaced along a major subhorizontal thrust fault (a horizontal fault below the surface). The cocorp data also indicate that the Brevard fault is a thrust that splayed, or

broke off, from the major subhorizontal thrust. Splaying may also have given rise to other faults in the area.

The cocorp study traces the horizontally layered reflections to under the Inner Piedmont and the Charlotte belt. Since these reflections are similar to and laterally continuous with the reflections under the Blue Ridge, we think the sedimentary layers are essentially continuous from the Valley and Ridge province to the Charlotte belt. Nevertheless, there is a substantial change in the character of the reflections at a distance of about 250 to 300 kilometers from the northwestern end of the profile. A series of eastward-dipping reflections, similar to those that often characterize the deposits on a continental slope, distinguish this part of the seismic section from the thin, layered reflection band to the west.

At the surface over the dipping reflections is the Elberton granite: a large body of igneous rock intruded into the crust. Under the folded and metamorphosed rocks of the eastern part of the Charlotte belt is another discrete, multilayered and horizontal sequence of reflections. The wedge of dipping reflections under both the Elberton granite and the Charlotte belt, together with horizontal reflections from under the belt, suggests there is a thick layered sequence of rocks between 12 and 18 kilometers below the surface.

Other significant changes in the reflection response of the deep crust are found at the southeastern end of the surveyed area. There anomalous horizontal reflections of between 10.5 and 11 seconds (equivalent to depths of between 30 and 33 kilometers) may correspond to the transition between the crust and the mantle. This transition is known as the Mohorovičić discontinuity. Reflections from the Mohorovičić discontinuity, and many other deep reflections as well, are not observed on the northwestern two-thirds of the profile. Other seismic data suggest that the crust under the Inner Piedmont and the Blue Ridge may deepen to between 40 and 45 kilometers. Over the Mohorovičić discontinuity at the southern end are intriguing westward-dipping reflections, which will be difficult to interpret until the profile is extended to the east. They could represent fault zones, parts of an ancient subduction-zone complex or perhaps even an earlier geological structure unrelated to the Appalachians.

It is still not known what the reflection changes at the transition between the Inner Piedmont and the

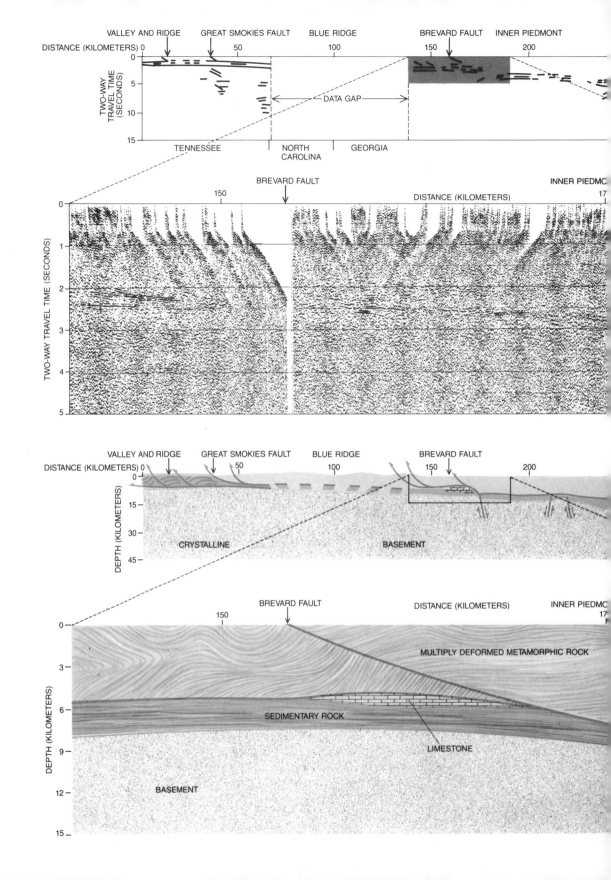

VALLEY AND RIDGE GREAT SMOKIES FAULT BLUE RIDGE BREVARD FAULT INNER PIEDMONT

DISTANCE (KILOMETERS) 0   50   100   150   200

TWO-WAY TRAVEL TIME (SECONDS)

DATA GAP

TENNESSEE   NORTH CAROLINA   GEORGIA

BREVARD FAULT

150   DISTANCE (KILOMETERS)   INNER PIEDMC 17

TWO-WAY TRAVEL TIME (SECONDS)

VALLEY AND RIDGE GREAT SMOKIES FAULT BLUE RIDGE BREVARD FAULT

DISTANCE (KILOMETERS) 0   50   100   150   200

DEPTH (KILOMETERS)

CRYSTALLINE   BASEMENT

BREVARD FAULT

150   DISTANCE (KILOMETERS)   INNER PIEDMC 17

MULTIPLY DEFORMED METAMORPHIC ROCK

SEDIMENTARY ROCK

LIMESTONE

DEPTH (KILOMETERS)

BASEMENT

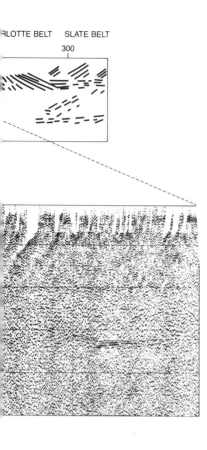

CHARLOTTE BELT   SLATE BELT
300

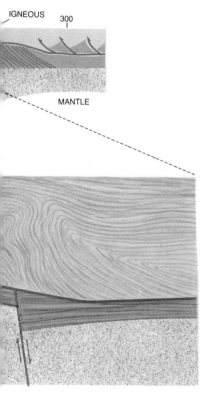

IGNEOUS   300

MANTLE

Figure 10.5 SEISMIC-REFLECTION PROFILE of the southern Appalachians is shown in a highly schematic form at the top. The vertical axis indicates the two-way travel times of the reflected waves. The horizontal axis locates the major geological features and indicates the distance along the surface from the northwestern end of the beginning of the COCORP-survey area. Below the schematic profile is a photograph of the actual profile across the Brevard fault. Next a possible geological interpretation of the profiles is shown. Crystalline rocks of chiefly continental origin are gray. Folded metamorphic rocks near the surface are wavy gray lines. Sedimentary strata are a light color. The inferred faults are lines with arrowheads that show the relative motion.

Charlotte belt represent. One interpretation has the eastward-dipping reflections under the Charlotte belt marking the points where the major thrust fault steepens and plunges deep into the crust. Because this interpretation cannot easily account for the horizontal reflections to the southeast under the Carolina slate belt we prefer a different one. We think the horizontal reflections and the thick sequence of eastward-dipping reflections under the Charlotte belt are caused by sedimentary rocks that were part of a continental margin and continental-shelf edge now buried under the overthrust sheet. These sedimentary rocks are now quite deep, and so they probably have been metamorphosed.

This interpretation can also explain the decrease in depth of the Mohorovičić discontinuity from northwest to southeast along the survey: it is the result of a transition from continental crust on the west to former oceanic crust or thinned continental crust on the east. This suggests that the deformed rocks of the Charlotte belt and the Carolina slate belt, like the deformed rocks of the Blue Ridge and the Inner Piedmont, were thrust over a pile of sediments. According to this interpretation, the thrusting must continue even farther to the east. Seismic studies now in progress should reveal whether or not it actually does.

If the reflections under the Charlotte and Carolina slate belts are indeed from sedimentary or metamorphosed-sedimentary layers, then before thrusting began the highly metamorphosed and deformed rocks of the Blue Ridge must have been a great distance away from their current position with respect to the continental interior. They were probably east of the current position of the Carolina slate belt. The distance of about 260 kilometers from the southeastern edge of the profiled area to the western edge of the Blue Ridge is the minimum lateral distance the rocks of the Blue Ridge were transported.

What does seismic-reflection profiling reveal about the tectonic history of the Appalachians? According to the theory of plate tectonics, the rigid plates of the lithosphere, the solid surface of the earth extending downward about 100 kilometers, float on the asthenosphere: a mobile and moderately fluid layer of the mantle a few hundred kilometers thick. Each plate consists of an upper layer of crust (between 35 and 40 kilometers thick for a continental plate and between five and 10 kilometers thick for an oceanic one) and a lower layer of solid and strong mantle. Oceanic lithosphere is created at the mid-ocean ridges, where magma continually wells up, cools and hardens to form the trailing edge of the moving plate. Such lithosphere is ultimately reabsorbed into the mantle in the subduction zones, the deep oceanic trenches created where two plates collide and one plunges under the other.

A subduction zone continues to consume lithosphere until a continent or an arc of oceanic islands impinges on it. Because the material of the continental crust is much lighter than that of the mantle most geologists think a continent can scarcely be subducted as a unit. If it indeed cannot, then the collision of a continent or an island arc with a subduction-zone complex will drastically change the nature of the subduction or even arrest it. The deformation of lithosphere at the collision site may give rise to a mountain range.

The southern Appalachians have evolved in a series of collisions of terranes, which are fragments of continental or island-arc material at the eastern edge of North America in the Taconic, the Acadian and the Alleghenian orogenies. Several models of the evolutionary process have been developed, and the COCORP data put major new constraints on them. We shall discuss one promising model that was developed by Hatcher to fit the surface geological data (see Figure 10.6). We have modified the model to take into account the COCORP subsurface information. The dates of the orogenic episodes have been determined from the deposition of sediments in the Appalachian basin and from the radioactive dating of metamorphic and igneous rocks. The overall composition of the colliding masses has been inferred from the types of rock in the various provinces and belts.

About 750 million years ago magma rising from the deep interior of the earth split a megacontinental expanse into at least two large continents (Laurentia, or proto–North America, and Gondwana, or proto-Africa) and at least two continental fragments that included the Inner Piedmont–Blue Ridge fragment and the Carolina slate belt fragment. In the period of early rifting the rocks that are now the volcanic and the metamorphosed sedimentary material of the Blue Ridge were deposited in the basin separating proto–north America from the Inner Piedmont–Blue Ridge fragment.

There is no indisputable evidence that the Piedmont and the Carolina slate belt fragments came from the same megacontinent that North America did. Nevertheless, the fact that some rocks in the Blue Ridge and the Piedmont have the same radioactive-dating age (about a billion years) as the basement rocks of eastern North America suggests that these crustal pieces underwent metamorphism at about the same time in the Precambrian. They may therefore have been part of the same continental block. The structure of the Appalachians is the result not of a single crustal block's colliding with North America but more likely of collisions of small continental fragments or island arcs in a way that resembles the numerous collisions occurring today in the southwestern Pacific as Australia moves toward Asia with a group of island arcs and continental fragments trapped between them.

Volcanism started in the island arc of the Carolina slate belt fragment some 650 million years ago. This means that subduction, which gave rise to the volcanic activity, also began at about that time. Some 500 million years ago the basin between proto–North America and the Inner Piedmont–Blue Ridge fragment began to close as a result of subduction. The very existence of sedimentary layers under the Blue Ridge and the Piedmont implies that the subduction zone dipped eastward; westward subduction at that time would probably have destroyed the sediments when volcanism took place. Before 450 million to 500 million years ago the sediments that gave rise to the sandstones and shales now found in the Valley and Ridge province came from proto–North America. For the next 250 million years sediments were derived chiefly from land in the opposite (easterly) direction. This shift in sedimentary history coincides with the first episode of deformation and metamorphism (the Taconic orogeny from 500 million to 450 million years ago) and can be attributed to the result of the closing of the basin between the Inner Piedmont–Blue Ridge fragment and proto–North America and the subsequent collision of these land masses. The sedimentary rocks of the east came from the thrust sheet that had started to move westward onto the continent.

It is not at all clear how a sheet between 10 and 20 kilometers thick was detached from the lower crust of the Inner Piedmont and the Blue Ridge and was then thrust over the continental shelf. Why did it split where it did and what became of the remaining 80 kilometers or so of underlying lithosphere? One hypothesis suggests that as the upper crust became detached the remaining lower crust and upper mantle continued to be subducted eastward

and collided with the island arc of the Carolina slate belt. Another similar hypothesis proposes that the polarity of the subduction in the basin reversed, so that a "flake" of the upper crust was pushed toward the continent as the lower crust of the Inner Piedmont and the Blue Ridge was subducted westward. (The flaking hypothesis had previously been adopted by E. R. Oxburgh of the University of Cambridge to explain some features of the Alps.)

The second episode of mountain building was the Acadian orogeny, from 400 million to 350 million years ago. Characterized by extensive metamorphism and deformation, the orogeny was triggered by the closing of the ocean basin between the Inner Piedmont–Blue Ridge fragment and the Carolina slate belt fragment. By that time the Inner Piedmont–Blue Ridge fragment was probably accreted to the proto–North American continent by overthrusting. Today the Kings Mountain belt may be the surface remnant of the ancient collision zone between the Carolina slate belt island arc and the Inner Piedmont. Although the size of the Carolina slate belt fragment is not known, the fragment could well have been quite wide because metamorphic rocks that are similar to the rocks of the Carolina slate belt extend under the coastal plain.

After the Acadian orogeny the next (and last) major compressional event in the southern Appalachians was the Alleghenian orogeny from 300 million to 250 million years ago. This mountain-building episode can be attributed to the collision of proto–North America and proto-Africa (or perhaps South America) to form the supercontinent of Pangaea. Although the Alleghenian orogeny was not as prominent in the northern Appalachians as the other two orogenies were, it gave rise in the southern Appalachians to large-scale overthrusting and extensive igneous activity.

At that time the Brevard zone broke through the surface, transporting carbonate sedimentary rocks from below. Radioactive dating indicates that many of the igneous bodies in the Piedmont were emplaced between 300 million and 250 million years ago. Large-scale overthrusting in this period has also been mapped in western Africa, although the western limit of the thrust faults has not yet been determined. We speculate that a segment of the African (or South American) continental shelf underthrust the eastern margin of the Carolina slate belt fragment, resulting in a fold-and-thrust belt that went in the opposite direction.

After the Alleghenian orogeny extensional tec-

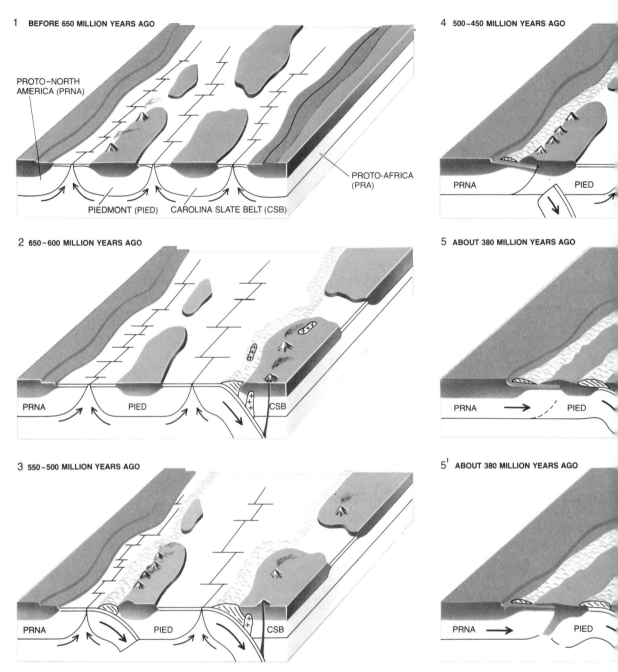

Figure 10.6 PLATE TECTONIC HISTORY of the southern Appalachians. Continental crust is dark gray, oceanic crust is white, and continental shelf rocks are light-colored. Initial rifting of North America away from Gondwana left smaller fragments, as labeled (*1*). Subduction gave rise to volcanic rocks of the Carolina slate belt (*2*). The closing of the ocean by collision of the Piedmont island arc (*3*, *4*) may have given rise to the Taconic orogeny. The thrusting over

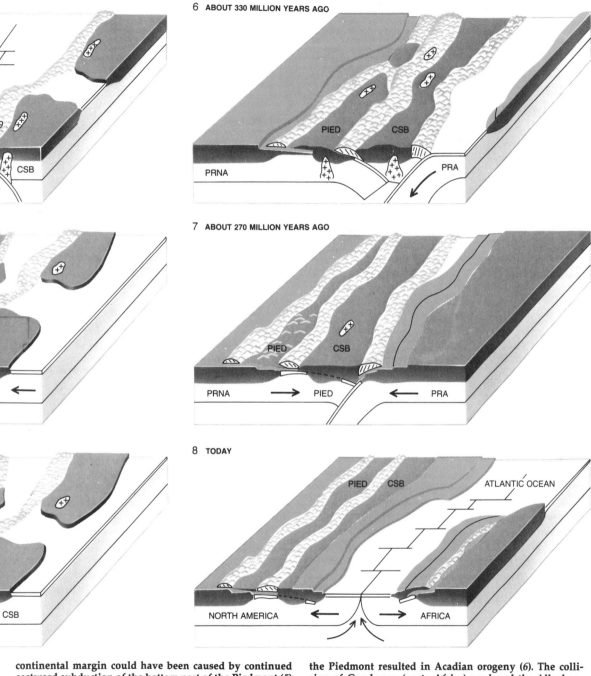

6 ABOUT 330 MILLION YEARS AGO

7 ABOUT 270 MILLION YEARS AGO

8 TODAY

continental margin could have been caused by continued eastward subduction of the bottom part of the Piedmont (5) or westward subduction of the bottom part of the Piedmont and Blue Ridge (5′). The collision of the Carolina slate with the Piedmont resulted in Acadian orogeny (6). The collision of Gondwana (proto-Africa) produced the Alleghenian orogeny (7). The opening of the Atlantic pushed the continents apart (8).

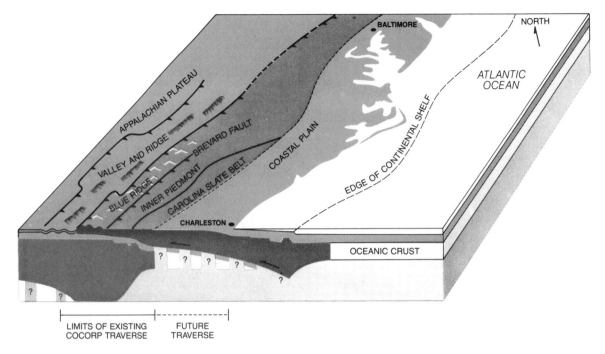

Figure 10.7 SUBSURFACE GEOLOGY OF THE EASTERN U.S. was inferred from the seismic-reflection profiles of the existing COCORP survey. The sedimentary strata are colored, the crystalline rocks of chiefly continental origin are dark gray and the basaltic oceanic crust is white. Between 500 million and 250 million years ago the continental shelf was overthrust by a thin near-horizontal sheet of crystalline rock. A new continental shelf has been forming off the Atlantic coast for 200 million years.

tonism again started to break a megacontinent (Pangaea) into smaller continents between 250 million and 200 million years ago. As the continents drifted apart the Atlantic Ocean was left in their wake. The Triassic basins of the eastern coast of the U.S., such as the grabens (troughs with near-vertical sides) in New Jersey and Connecticut, formed at that time. As the Atlantic grew, the current continental shelf was built up off the eastern coast of North America (and off the western coast of Africa and the northern coast of South America) (see Figure 10.7).

Is it possible that large-scale thin thrust sheets are a general result of continental collisions? There are several indications that they are. We have discussed the fold-and-thrust belts of western Africa and northern South America, which have a geological structure quite similar to that of the Appalachian belts. David Gee of the Geological Survey of Sweden has mapped major horizontal thrust faults in the Caledonide mountains of Scandinavia, some of which seem to have been displaced by hundreds of kilometers.

The kind of thin-skinned thrusting found in the Valley and Ridge province can also be seen in the fold-and-thrust belts of the Montana and Alberta cordillera, which is part of the Rockies. This area is a future site for COCORP studies that will try to determine the western limit of the thrusting. Thin-skinned thrusting may also have been dominant in other mountain ranges, including the Alps, the Himalayas and the Zagros of the Middle East. Some geologists believe the kind of overthrusting evident in the Appalachians is even now taking place in the Timor region of the southwestern Pacific, where the Australian continental shelf has underthrust the Timor island arc to the northwest for perhaps 150 kilometers (see Figure 10.8).

The entrapment of sedimentary rocks under large metamorphic sheets has major consequences not only for the growth of continents but

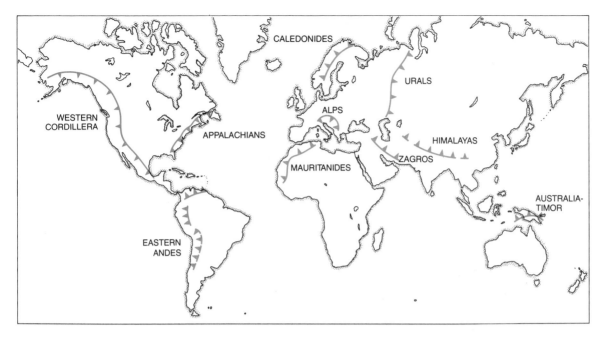

Figure 10.8 MOUNTAIN RANGES AROUND THE WORLD (the western Cordillera, the eastern Andes, the Mauritanides, the Alps, the Urals, the Caledonides, the Zagros, the Himalayas and Australia-Timor) may consist of thin thrust sheets as do the southern Appalachians.

also for the recovery of gas and oil. The discovery of sedimentary material under the metamorphic rocks of the Blue Ridge and Piedmont provinces calls for reevaluating the possibility of extracting hydrocarbons from that part of the Appalachians. Perhaps the metamorphism of the sediments has not been intense enough to remove all the hydrocarbon deposits from the rocks under the thrust sheet.

The discovery of the layer of sediments under the core of the Appalachians has implications that go far beyond this particular mountain range. The mechanism of thin-skinned thrusting involving basement rocks may have been responsible for the formation of mountain belts for as long as plate-tectonic processes have been operating. In that case the evolution of continents is characterized not by lateral accretion along vertical boundaries but by the juggling and stacking of thrust sheets. Much work remains to be done, but we suspect that for at least half of the earth's history such thrusting could well have been a chief mechanism in the evolution and growth of continents.

The Growth of Western North America

Over the past 200 million years the continent has been extended westward by repeated collisions with smaller land masses, some of which appear to have come from thousands of kilometers away.

· · ·

David L. Jones, Allan Cox, Peter Coney and Myrl Beck
November, 1982

According to the theory of plate tectonics, the earth's land masses are riding on great plates of the earth's crust that are in continuous motion with respect to one another. Since the theory became widely accepted less than two decades ago earth scientists have generally believed continents grow slowly and steadily, somewhat like trees, accumulating rings along their outer margins. Such growth rings consist of several different kinds of rock. Some are ocean-floor rocks scraped off against the edge of a continent where an approaching plate has plunged under the continental plate, the process known as subduction. Some of the rocks are derived from volcanic arcs, the chains of volcanic islands that form above subduction zones. Many of the rocks originate as sediments deposited on the continental shelf by rivers.

It now seems, however, that the growth of continents is not slow and steady. New evidence shows that it has been episodic, with the last major pulse in the growth of North America beginning no more than 200 million years ago. Virtually the entire Pacific Coast from Baja California in the south to the tip of Alaska in the north, and extending inland an average distance of some 500 kilometers, was grafted onto the preexisting continent by the piecemeal addition of large, prefabricated blocks of crust, most of which were carried thousands of kilometers east and north from their sites of origin in the Pacific basin. The horizontal dimensions of the individual blocks ranged from hundreds to thousands of kilometers.

Many of the blocks are of oceanic origin, consisting of oceanic crust, islands, plateaus, ridges or island arcs. A few blocks are clearly fragments of other continents. Some have traveled several thousand kilometers with remarkably little internal deformation. After they made contact with North America the blocks were usually sliced by shear faults and drawn out into thin strips parallel to the continental margin. During and after collision they were in many instances rotated. Thus western North America is a collage of accreted blocks that has been shaped to its present configuration over the past 200 million years by the impact of oceanic plates, each block carrying a burden of exotic rocks. The process by which the edge of a continent is modified by the transport, accretion and rotation of large crustal blocks is now often called microplate tectonics. The blocks themselves are called terranes.

Microplate tectonics is a significant addition to that part of plate tectonics which describes the interactions of plates along continental margins that are called active. The basic theory of plate tectonics recognizes two ways continental margins can grow seaward. Where two plates such as the African plate and the South American plate are moving away from a midocean rift that separates them, the continental margins on those plates are said to be passive, or rifted. Such continental margins grow slowly from the accumulation of riverborne sediments and of the carbonate skeletons of marine organisms, which are deposited as limestone. Suites —unbroken sequences—of such accretions, consisting of nearly flat strata, are called miogeoclinal deposits. Since most miogeoclinal deposits are undeformed and exhibit an unbroken history, it is evident that passive margins are generally not associated with mountain building.

A long active, or convergent, margins, such as those that ring most of the Pacific basin, continents tend to grow much faster. At an active margin an oceanic plate plunges under a continental plate, with the continental plate scraping off deep-ocean sediments and fragments of basaltic crust that then adhere to the continental margin. Simultaneously the plate plunging under the continental margin heats up and partially melts, triggering extensive volcanism and mountain building. A classic example is the Andes of the west coast of South America.

In the original plate-tectonic model western North America was described as being a passive margin through the late Paleozoic and early Mesozoic eras (roughly 350 to 210 million years ago), after which it became an active margin. It was assumed that the continent grew to a limited extent along this margin as sedimentary and igneous rocks of oceanic origin were accreted in a few places, as in the Coast Ranges of California. The model was successful in explaining such disparate features as the Franciscan rocks of the California Coast Ranges, formed by local subduction processes, and the granitic rocks of the Sierra Nevada, farther to the east, which clearly originated as the roots of volcanoes similar to those of the Andes.

The basic plate-tectonic reconstruction of the geologic history of western North America remains unchanged in the light of microplate tectonics, but the details are radically changed. It is now clear that much more crust was added to North America in the Mesozoic era (248 to 65 million years ago) than can be accounted for by volcanism along island arcs and by the simple accretion of sediments from the ocean floor. It has also become evident that some terranes lying side by side today are not genetically related, as would be expected from simple plate tectonics, but almost certainly have traveled great distances from entirely different parts of the world.

Here we shall address four major questions. How can one recognize the separate terranes that were accreted to form the tectonic collage of western North America? How can one establish where the terranes originated and how far they have traveled? What are the structural relations between accreted terranes? How do the terranes become accreted to the growing edge of the continent?

Answers to these questions call for close collaboration among earth scientists from many subdisciplines. For example, geologists, geophysicists and paleontologists all have their own methods for recognizing pieces of the earth's crust that have been transported to their present location from distant locations. To take a simple but realistic example, Baja California and the narrow slice of California that lies west of the San Andreas fault are sliding northward at the rate of about five centimeters per year in relation to the rest of North America. Fifty million years from now, if the movement continues, the California rocks will have accreted along the continental margin of Alaska.

The discontinuity between the "native" Alaskan rocks and the "foreign" rocks of California would reveal itself in three principal ways. First, there would be abrupt discontinuities in rock sequences across major faults, implying very different geologic histories in terranes that had become adjacent. Second, there would be similar discontinuities in the fossils of plants and animals; tropical forms in the displaced rocks could readily be distinguished from the cool-temperate forms in the native Alaskan rocks. Third, the two kinds of rocks would exhibit markedly different magnetic characteristics. When molten rock cools, its intrinsic magnetism is aligned with the earth's local magnetic field. Thus rocks formed near the Equator, where the lines of force in the earth's magnetic field are nearly horizontal, would show a shallow paleomagnetic inclination. Native Alaskan rocks, which solidified at high latitudes where the lines of force in the earth's field plunge downward, would show a steep paleomagnetic inclination.

The original recognition of exotic terranes in western North America came about through observations of the first two kinds of anomalies: discontinuities in geology and in paleobiology. The explanation that these anomalies must be due to enormous displacements of large crustal blocks was advanced in paleomagnetism.

Figure 11.1 shows the distribution of major terranes in western North America. Many smaller terranes have been identified, but they cannot be shown at this scale. Each terrane represents a separate geologic entity characterized by a distinctive sequence of rocks that differs markedly from the sequence found in neighboring rocks. Each terrane is bounded on all sides by major faults; transitional strata or rocks that would serve to link one terrane to another are missing.

The defining characteristic of a terrane is a unique sequence of geologic events. The events include the deposition of volcanic and sedimentary rocks, the intrusion of granitic rocks and earth movements such as folding and faulting. The formation of ore deposits may also be part of the geologic history; one important result of the study of terranes has been an understanding of why certain mineralization processs stop abruptly at what are now recognized as terrane boundaries.

One of the first terranes to be identified is the Cache Creek terrane of British Columbia. As early as 1950 M. L. Thompson and Harry E. Wheeler of the University of Washington and W. K. Danner of Wooster College pointed out that certain distinctive marine microfossils known as fusulinids, dating back to the Permian period about 250 to 290 million years ago, are widely distributed in westernmost North America but are totally unlike species found farther east in the Rockies and in the middle of the continent (see Figure 11.2). The western forms belong to species widely distributed through China, Japan, the East Indies and the Malay Peninsula. The Asian fusulinids help to define the Tethyan faunal province, a term that alludes to the ancient Tethys Ocean that lay southeast of the Eurasian land mass (see Figure 11.3). The species of fusulinids found throughout Nevada, Texas and Kansas belong to the North American faunal realm (see Figure 11.4).

Early investigators hypothesized that the exotic Tethyan fusulinids reached western North America by migrating through a complex system of narrow "seaways" that somehow allowed travel from west to east but not the reverse. The seaways are the marine analogues of the "land bridges" invoked before plate tectonics to explain the similarly puzzling distribution of land animals. In 1968 J. Tuzo Wilson of the University of Toronto, an early proponent of plate tectonics, suggested that the presence of anomalous marine fossils in North America could be explained if the Pacific Ocean had once been closed, so that Asia and North America were in contact. On the reopening of the Pacific, scraps of Asia with Tethyan fossils would be left behind, plastered to the rifted margin of North America. Such a sequence of the closing and opening of a major ocean basin, now known as the Wilson cycle, is well documented for the Atlantic. There is little or no evidence, however, to support a complete closure of the Pacific basin, at least not over the past several hundred million years.

In 1971 James W. H. Monger of the Canadian Geological Survey and Charles A. Ross of Western Washington University advanced the simple hypothesis that the Tethyan Permian fusulinids and the rocks in which they are found formed during the Permian period near the Equator as part of the ocean floor. The rocks were later transported northward to Canada on an oceanic plate and eventually added to North America by accretion.

In order to establish the origin of the Cache Creek terrane of British Columbia it is important to know whether the equatorial marine organisms of the Tethyan fauna were confined to a single equatorial province or whether they were distributed along the entire Permian Equator. When the present distribution of Tethyan fusulinids is plotted, it seems quite clear that those found in a belt extending from the Mediterranean on the west to Borneo and possibly Japan on the east are indigenous. Their ancestral waters were the Tethys, which in Permian time lay

Figure 11.1 AREAS ADDED TO WESTERN NORTH AMERICA over the past 200 million years. Rocks in these terranes differ sharply in geology, paleontology and paleomagnetic properties from rocks in the ancient North American craton: the primitive continent (*light gray*). Many of the terranes, including all of those in color (with the possible exception of Yukon-Tanana), are made up of rocks that originally formed on the ocean floor. Some of the terranes embody paleomagnetic evidence that they originated thousands of kilometers to the south of their present position. The barbed line near the western edge of the ancient continent marks the eastern limit of the Laramide orogeny, mountain building that began 150 million years ago and ended 50 million years ago.

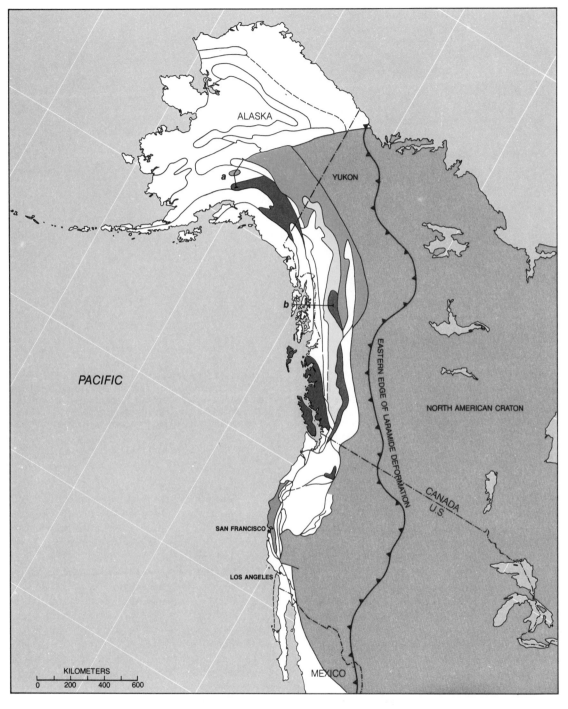

ALASKA

a

YUKON

b

PACIFIC

EASTERN EDGE OF LARAMIDE DEFORMATION

NORTH AMERICAN CRATON

CANADA
U.S.

SAN FRANCISCO

LOS ANGELES

MEXICO

KILOMETERS
0   200   400   600

ACCRETED TERRANES

NAME	TYPE
CHULITNA	OCEANIC BASIC
CACHE CREEK	OCEANIC BASIN–CARBONATE PLATEAU
FRANCISCAN	DISRUPTED OCEANIC BASIN
STIKINE	VOLCANIC ISLAND ARC
WRANGELLIA	VOLCANIC ARC – OCEANIC PLATEAU
YUKON-TANANA	METAMORPHIC

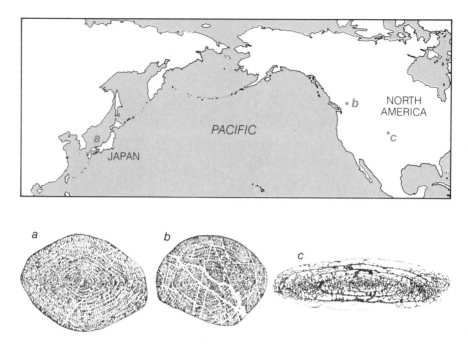

a   b   c

Figure 11.2 FOSSIL EVIDENCE FOR THE TRANSLOCA-TION OF TERRANES is supplied by differences in fusu-linids, marine microfossils found in rocks from widely separated sites. The micro-fossils date back to the Permian period, 240 to 290 million years ago. The similar microfos-sils at sites *a* and *b* are described as Tethyan fusulinids because both presumably are from the ancient Tethys Ocean (see Figure 11.3). They can be distinguished from North American fusulinids (*c*) in external form and inter-nal structure. Tethyan fusulinids in western North Amer-ica were evidently carried on terranes that traveled thou-sands of kilometers.

between India, Tibet, Australia and Africa to the south and Europe and Asia to the north.

The Tethys embraced at least five substantial is-land masses that subsequently became accreted to the eastern margin of Asia at locations distributed from near the present Equator to the present Bering Sea. These accreted regions all have Tethyan fusu-linids. Equally important is the observation that there are no Tethyan fusulinids in equatorial Per-mian rocks indigenous to the Western Hemisphere. The reason for the present dispersal of Tethyan fu-sulinids is clearly not a faunal migration during Per-mian time but rather a subsequent displacement of the Permian terranes that originally formed in the Tethys basin and on which the fusulinids were de-posited and fossilized.

This simple explanation for the origin of the enig-matic fossils of the Cache Creek terrane carries an important implication. The exotic-fusulinid rocks are found 500 kilometers inland from the coast of North America. If the rocks are indeed of exotic origin, as Monger and Ross speculated, one must suspect that the rocks lying seaward of them to the west are exotic too. This suspicion is now being amply confirmed.

Many of the Paleozoic and Mesozoic rocks with ages between 590 and 65 million years found in parts of Alaska, British Columbia, Washington, Or-egon, western Nevada, California and western Mexico lack obvious connections with the ancient craton, or core, of North America. The ancient west-ern edge of North America at the last epoch when the continental margin was still a passive one can be delineated with reasonable accuracy on the basis of both lithological (rock type) and geochemical cri-teria. A major characteristic of such a margin is the presence of basement rocks under thick Paleozoic sedimentary rocks derived from continental sources and deposited in deep water. In contrast to this sequence, the terranes found west of the ancient

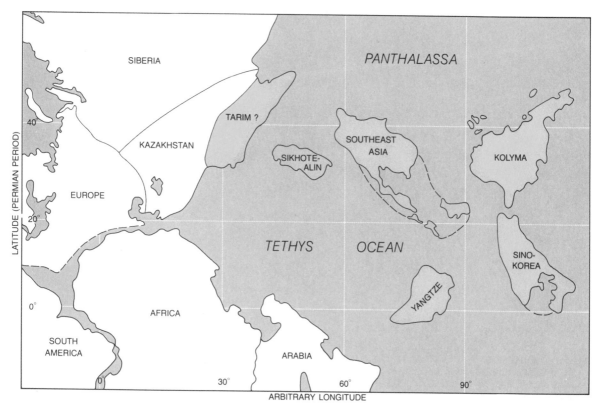

Figure 11.3 RECONSTRUCTION OF TETHYAN REGION at a time 250 million years ago in the Permian period shows the possible disposition of various blocks that were subsequently accreted to the Eurasian continent. The Tethyan region constituted a unique equatorial faunal prov- ince, for which the fusulinids are a key fossil index. The present location of the blocks is depicted in Figure 11.4. The two maps are based on studies by M. W. McElhinny, B. J. J. Embleton, X. H. Ma and Z. K. Zhang.

edge of North America consist of rocks characteristic of island arcs, oceanic crust and of sediments derived from them.

A key geochemical marker of the boundary between ancient continental crust and exotic rocks of oceanic affinity is a change in the ratio of two isotopes of strontium: strontium 87 and strontium 86. In ancient continental crust of Precambrian age (older than 590 million years) the ratio of strontium 87 to strontium 86 is high and in oceanic crust it is low. Such a distinctive change in isotope ratio provides a marker that coincides well with the lithological discontinuity.

The edge of the ancient continent defined by the two coincident markers lies from a few hundred to many hundreds of kilometers east of the present continental margin. This implies that all the rocks lying west of the ancient continental edge have been added by some accretionary process. Most of the accretion took place in a relatively brief period extending from about 200 million years ago to about 50 million years ago. We now believe that in this 150-million-year period exotic fragments derived from unknown sources in the Pacific were swept against and added to the western edge of North America. More than 100 highly diverse fragments have now been identified.

Terranes can be conveniently divided into four general categories: stratified, disrupted, metamorphic and composite. Stratified terranes are characterized by coherent stratigraphic sequences in

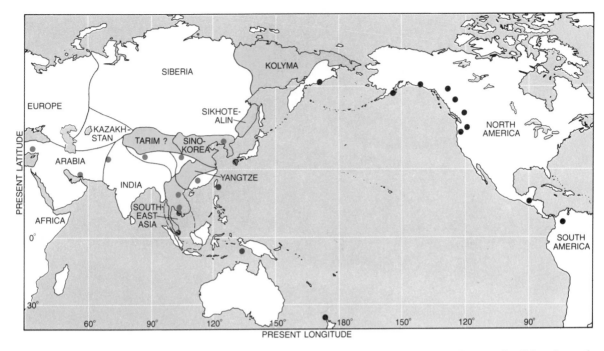

Figure 11.4 PRESENT DISTRIBUTION OF TETHYAN FUSULINIDS within their home territory is indicated by colored dots. Their presence in eastern Siberia, New Zealand and in the Western Hemisphere (*black dots*) provides strong evidence for large-scale tectonic dislocations of crustal blocks formerly in the Tethyan region. Tethyan blocks that joined Eurasia are shown in the locations proposed by M. W. McElhinny.

which the order of deposition between successive lithologic units can be demonstrated. Basement rocks may or may not be preserved. Rock sequences within stratified terranes can be subdivided into three broad subcategories depending on whether the origin of the rocks is predominantly continental crust, oceanic crust or volcanic island arc. If the terranes have had a complex tectonic history, the strata will exhibit a succession of these crustal types.

Fragments of continents, the first subcategory, are characterized by the presence of a Precambrian basement with an overlying sequence of shallow-water sediments of Paleozoic and Mesozoic age. Included in this subcategory are sedimentary rocks of continental affinity that have become detached from their basement substratum.

Fragments of oceanic crust, the second subcategory, are characterized by sequences of extruded molten rock typical of oceanic crust, usually overlain with layers of siliceous chert composed mainly of the skeletons of radiolarians (marine protozoans). Included in this subcategory are deep-sea deposits

that have become detached from their basement substratum.

Fragments of volcanic arcs, the third subcategory, consist of stratified terranes composed chiefly of volcanic rocks of the plutonic (deep igneous) roots of volcanic island arcs, together with sedimentary debris originating fom volcanos. The rocks in this subcategory are similar in composition to those of currently active volcanic arcs such as the Aleutians.

The second general category, disrupted terranes, is represented by blocks of heterogeneous lithology and age, usually set in a matrix of sheared shale or serpentinite (a rock poor in silica but rich in iron and magnesium). Most of these terranes harbor fragments of oceanic crust, blocks of shallow-water limestone, deep-water chert and packages of graywacke ("dirty sandstone") incorporating lenses of conglomerate. Many disrupted terranes also include blue schists (metamorphic rock formed under high pressure), which can be either indigenous or exotic.

Composite terranes, the third general category, are assembled from two or more distinct terranes that were amalgamated and then shared a common geologic history before their accretion to North America. The fourth category, metamorphic terranes, consists of rocks that have been subjected to terrane-wide geologic changes before or after being accreted to North America, including the development of metamorphic minerals to such a degree that the original stratigraphic features and relations are obscured.

Terranes vary enormously in size. Some cover tens of thousands of square kilometers, others only a few hundred. Many terranes that arrived in one piece were subsequently broken up and now consist of separate patches that can be correlated stratigraphically.

The remarkable fact that emerges from the analysis of the various terranes is that each terrane records a geologic history significantly different from its neighbors'. In most instances the differences are so pronounced that it would be inconceivable for the rocks of neighboring terranes to have formed in close proximity. The differences between terranes become apparent when the geology of three terranes in southern Alaska—Wrangellia, Chulitna, and Cache Creek—are compared with the stable western interior margin of North America in northeastern British Columbia (see Figures 11.1 and 11.5). The North American rocks (left column of Figure 11.5) that were laid down from 570 million years to about 200 million years ago consist of interbedded sandstone, shale and conglomerate and subordinate limestone deposited in shallow water. Such sediments are characteristic of passive continental margins, and they indicate that over this entire period the northwestern margin of North America was slowly subsiding, as the Atlantic margin is today. About 200 million years ago the margin became active as oceanic plates began to subduct beneath the margin.

The oldest rocks in Wrangellia are volcanic rocks characteristic of island arcs, capped by a thin sequence of shallow-water marine shale (see Figure 11.6). Limestones interbedded with the shale have fusulinids of Permian age that are entirely distinct from the Tethyan forms in the neighboring Cache Creek terrane. Overlying these fossil-bearing sediments is a thick sequence of volcanic basalt, possibly formed in a rift or an oceanic plateau reminiscent of the Ontong Java plateau (see Figure 7.6). The first lava flows formed below sea level, but the volcanic pile soon rose above sea level, ultimately

amounting to 100,000 to 200,000 cubic kilometers of basalt. Volcanism ceased in late Triassic time and the entire plateau subsided below sea level. The first sedimentary rocks were shallow-water carbonate deposits similar to those forming today in the warm waters of the Persian Gulf. With progressive sinking the carbonate deposits were gradually overlain by deep-water deposits in which remains of deep-water fauna are abundant. There is no detritus of continental origin in these post-volcanic sediments. We suspect that at this time Wrangellia was isolated in midocean, probably near the Equator. Its long northward drift had begun. The earliest continental sediments were deposited in Cretaceous time as Wrangellia began to collide with North America.

The Cache Creek rocks consist of a sequence of Paleozoic shallow-water limstones several thousand meters thick deposited directly on oceanic crust (now ophiolite) approximately 350 million years in age. The limestones contain Tethyan fusulinids of Permain age. In Triassic time the limestone platform subsided further and was overlain by deep-water deposits. In Jurassic time, 80 million years ago, the Cache Creek terrane completed its voyage from the western Pacific and docked in British Columbia.

Barely 50 kilometers long, the Chulitna terrane records a long and complex history of oceanic and continental sedimentation that is unique in North American geology (see Figure 11.7). The oldest rocks are a Devonian-Mississippian ophiolite complex (labeled "oceanic crust" in Figure 11.3), overlain by rocks formed from deep-water pelagic sediments. Late Paleozoic rocks and the earliest Mesozoic rocks include conglomerates derived from an island arc, together with shallow-water carbonate rocks. Because there is no continental detritus in this sequence it must have formed in a midoceanic setting. Conditions changed abruptly in the Triassic (243 to 213 million years ago) with the sudden influx of coarse, quartz-rich detritus (forming the red beds of Figure 11.7) mixed with fragments from the terrane's own oceanic basement. These deposits may record the docking of the Chulitna terrane against North America in low latitude.

All this evidence shows that the Chulitna terrane underwent a profound change in Triassic time from having a strictly oceanic setting to being incorporated into a continental margin. Intense folding, faulting and uplift in the course of the collision led to the erosion of the oceanic basement of the terrane and to the mixing of the detritus with

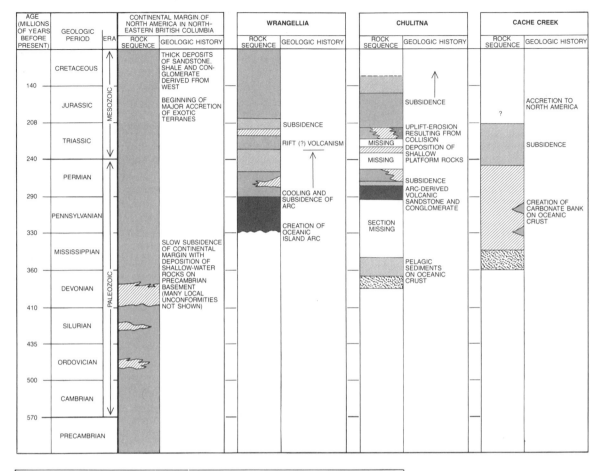

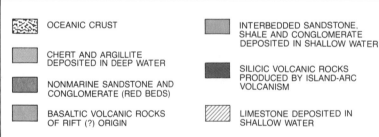

Figure 11.5 SCHEMATIC STRATI-GRAPHIC SECTIONS illustrate the differences in geologic history between the continental margin of North America in northeastern British Columbia, and the exotic terranes of Wrangellia, Chulitna and Cache Creek (see Figure 11.1 for locations.)

material from the adjacent continent. None of these dramatic events is recorded in the nearby rocks of Wrangellia. Although the two terranes are now next-door neighbors, they have completely different histories.

A major unsolved puzzle is where the Chulitna terrane was formed. In the stratigraphic and tectonic record it is totally unlike any other terrane known in North America. An origin far to the south is indicated by two independent lines of evidence. First, Chulitna has thick piles of reddish Triassic sediments ("red beds") whose only counterparts are found almost exclusively far to the south, below the U.S.-Canada border. Second, Triassic fossils in and below the Chulitna red beds are similar to forms known only from southern latitudes.

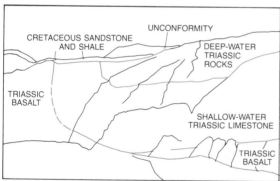

Figure 11.6 CLIFF IN THE WRANGELL MOUNTAINS some 400 kilometers east of Anchorage, Alaska, displays a 100-million-year history of the Wrangellia terrane. The principal geologic features visible in the photograph are depicted in Figure 11.7. The oldest rocks, dating back to the Triassic period about 240 million years ago, were formed on an island arc somewhere in the ancient Pacific. The finely bedded limestone deposits, some 1,200 meters thick, hold the skeletons of shallow-water marine organisms that settled on the basaltic platform as it slowly subsided during the Triassic. The limestone was then covered by sediments composed mainly of deep-water organisms such as sponges and radiolarians. The ridge at the left, consisting of shallow-water sandstones and shales, was deposited in the Cretaceous some 120 million years ago.

The remarkable geologic differences between neighboring Chulitna and Wrangellia are only two of many examples that could be cited. The essential point is that each terrane records a unique sequence of historical events that cannot be duplicated in detail anywhere else in North America. The discrimination of terranes is reasonably objective; it is based entirely on the preserved geologic data. The significance of the differences between the individual terranes and between the terranes and the an-

cient continent of North America, however, is the subject of continuing analysis and interpretation. The outstanding questions are: Where did the terranes originate? When and by what path did they move? In response to these questions paleomagnetic studies are now yielding significant new information.

They key to measuring the movement of terranes is close analysis of the magnetism frozen into basaltic and other igneous rocks at the time they solidi-

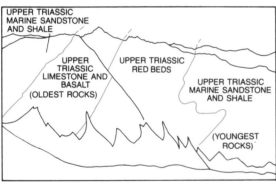

UPPER TRIASSIC
MARINE SANDSTONE
AND SHALE

UPPER
TRIASSIC
LIMESTONE AND
BASALT
(OLDEST ROCKS)

UPPER TRIASSIC
RED BEDS

UPPER TRIASSIC
MARINE SANDSTONE
AND SHALE

(YOUNGEST
ROCKS)

Figure 11.7 EXPOSED STRATA OF CHULITNA TER-
RANE lie about 60 kilometers to the northwest of the
northern boundary of the Wrangellia terrane. The terrane
exhibits a distinctive suite of rocks found nowhere else in
Alaska or farther south in North America. The section in
the photograph is folded and overturned so that the oldest
rocks are the light- and dark-banded rocks at the left.
These are of upper (late) Triassic age; the light strata are
basalt, the dark strata limestone. They are overlain deposi-
tionally at the right by upper Traissic red beds (sandstone
and conglomerate). The brown rocks still farther to the
right are shallow-water sandstone and shale harboring an
abundance of shallow-water marine fossils from the end of
the Triassic. Strata were presumably overturned when
Chulitna terrane reached Alaska 90 million years ago. Vertical relief is 600 meters.

fied from the molten state. As we have indicated, the inclination of the magnetic vector locked into the rocks at the time of their formation is more or less horizontal near the Equator and gets steeper with distance north or south of the Equator. The orientation of the magnetic vector is also described by a second value: the declination, or the angle between the vector and true north.

The paleomagnetic inclination reveals how far the rocks were from the geographic North Pole when they were formed (see Figure 11.8). The distance is derived from a simple equation based on the assumption that the geomagnetic field can be represented by a magnetic dipole, or bar magnet, aligned with the earth's axis of rotation. At any one instant this is not exactly true because the geomagnetic field shows considerable variation. The dipole assumption is nonetheless valid if one gets an average inclination from rock strata whose ages span a time interval of at least several tens of thousands of years. Paleomagnetic data can then establish the latitude at which rocks first formed with an accuracy of about five degrees.

The second value, the paleomagnetic declination, establishes a paleonorth direction that points toward the ancient geographic and magnetic pole. As with inclination, meaningful values for declination call for the averaging of paleomagnetic directions in rocks spanning a substantial period of time. The accuracy of determining declination varies with the original latitude of formation, being greatest for rocks formed near the ancient Equator.

Having found the paleomagnetic inclination and declination at some site for rocks of a given age, finding the pole is a straightforward exercise in spherical geometry. The declination tells one that when the rocks were formed, the ancient pole lay along a great circle passing through the sampling site and deviating from the present true north direction by a certain number of degrees equal to the declination. The inclination tells one the distance to the ancient pole at the time the rocks solidified. One assumes that in the distant past the mean paleo-

magnetic pole and the ancient geographic pole coincided, just as they have in the more recent past. Because the North American plate has been moving in relation to the earth's axis of rotation the pole as seen from North America appears to have moved.

Here is an example of how paleomagnetism can reveal whether or not a terrane of, say, Triassic age is displaced with respect to a stable North America. One begins by determining the mean location of the Triassic paleomagnetic pole with rocks obtained from the stable part of the continent. One then determines the paleomagnetic inclination of Triassic rocks in the terrane of interest. This measurement establishes the terrane's paleolatitude, or distance from the pole, in the Triassic. The terrane therefore must have been somewhere on a circle of that paleolatitude centered on the mean Triassic paleopole for the stable continent. If the circle happens to pass through the present-day sampling site, the terrane is not displaced except for a possible movement along the circle.

Studies of this type made in western Canada by Raymond W. Yole of Carleton University and Edward Irving of the Canadian Department of Energy, Mines and Resources, in Alaska by Duane R. Packer and David B. Stone of the University of Alaska and

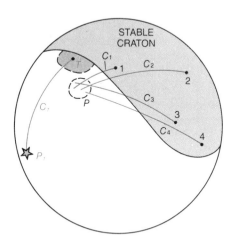
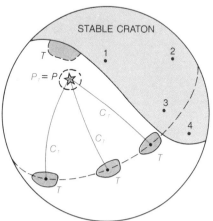

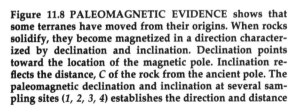

Figure 11.8 PALEOMAGNETIC EVIDENCE shows that some terranes have moved from their origins. When rocks solidify, they become magnetized in a direction characterized by declination and inclination. Declination points toward the location of the magnetic pole. Inclination reflects the distance, C of the rock from the ancient pole. The paleomagnetic declination and inclination at several sampling sites (1, 2, 3, 4) establishes the direction and distance and hence the location of the mean cratonal pole, P (*left*). Rocks of the same age from the possible terrane, T, are analyzed to obtain a paleomagnetic pole P_T for the terrane (*star at left*). At the time the rocks formed the location of P_T must have coincided with P (*right*). It is evident that the terrane must have originated at some location along the small circle of radius C_T.

J. W. Hillhouse of the U.S. Geological Survey and in Washington, Oregon and California by one of us (Beck) have established that many of the terranes in western North America have traveled northward thousands of kilometers. The paleomagnetic results are particularly striking for Wrangellia (see Figure 11.9). Rocks obtained from parts of the Wrangellia terrane on Vancouver Island in British Columbia and from the Wrangell Mountains in Alaska show that the rocks at both sampling sites, now 2,500 kilometers apart, originally formed near the Triassic Equator at essentially the same latitude. Their present separation is the result of strike-slip faulting during and after their accretion to North America,

which appears to have strung Wrangellia out in a north-south direction.

Another surprising result of the paleomagnetic studies has come from the determination of paleomagnetic declinations. It has been discovered that many of the terranes in western North America have rotated, most of them in a clockwise sense and some in excess of 70 degrees (see Figure 11.10). In some of these terranes the rotations call for a rethinking of the local geology. For example, in the Coast Range of Oregon marine sediments laid down in the Eocene epoch reveal the direction of currents on the ocean bottom. In the absence of paleomagnetic information it was believed the currents

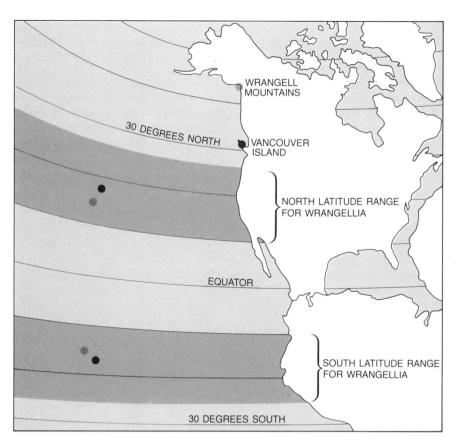

Figure 11.9 ORIGIN OF THE WRANGELLIA TERRANE has been narrowed down on the basis of paleomagnetic evidence to one of two probable regions by R. W. Yole and E. Irving: the Wrangell Mountains and Vancouver Island. In the late Triassic the rocks of both sites were formed as part of an island in the proto-Pacific some 16 degrees either north or south of the Traissic Equator. Allowing for possi-

ble sources of error, the paleomagnetic data establish that Wrangellia was situated in one of the two shaded bands. Whether the origin of the terrane was north or south of the Equator depends on whether the magnetism of the Triassic rocks was "frozen" when the polarity of the earth's magnetic field was "normal" (what it is today) or when it was reversed.

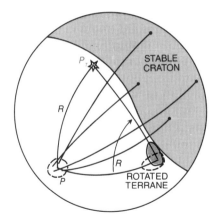

Figure 11.10 WHEN A TERRANE RO-
TATES by some angle, R, at some time
after the magnetism of its rocks was frozen, the apparent position deduced for
the paleomagnetic pole is similarly rotated. On the basis of data from the
stable craton the pole is known to have
been at location P. The rotation of the
terrane makes it appear that the pole was
at P_T rather than P.

flowed in a northerly direction parallel to the present continental margins. Paleomagnetic studies by Robert W. Simpson of the U.S. Geological Survey and one of us (Cox) show, however, that the rocks that recorded the bottom currents have been rotated clockwise by more than 50 degrees since the time of their formation, so that the true direction of Eocene bottom currents was northwesterly, in a direction away from the coast.

In a terrane that has not been rotated the paleomagnetic declination will point toward the paleomagnetic pole as it is determined from rocks of the same age on the undisrupted part of the continent. If a terrane has been rotated, the declination of its rocks will not agree with the mean declination determined from the stable part of the continent. From studies in Washington, Oregon and California one of us (Beck) concluded in 1976 that many terranes have rotted in a clockwise direction.

Rotations are found in terranes both with and without large latitudinal displacement. Rottions of the terranes that have been displaced can reasonably be attributed to changes in orientation in the course of the terrane's displacement and docking. The rotations of the terranes that have been only slightly displaced are less easily explained. We shall give two examples.

In southern California, Bruce P. Luyendyk and Marc J. Kamerling of the University of California at Santa Barbara have measured clockwise rotations of more than 60 degrees in rocks only 13 million years old (see Figure 11.11). What tectonic forces could have produced rotations at the rate of nearly five degrees per million years? The ultimate cause must

certainly be deformation produced by the north-westerly motion of the Pacific plate past North America. The sense of this motion is described as dextral, or right-handed, since an observer on either plate would see the other plate moving to the right. The challenge lies in discovering the precise mechanism by which dextral shear across the San Andreas fault creates the observed clockwise rotation.

In simple plate tectonics all displacement between two plates is assumed to occur along a single fault (see Figure 11.12a,b). Therefore if one were to paint a straight line across a boundary such as the San Andreas fault, after a million years the line would be offset by 50 kilometers or so, with the line segments on both plates remaining straight and parallel to each other. There would be no rotation. Similarly, if the plate motion were to be accomplished by a series of parallel faults, the painted line would simply be offset by a series of parallel steps, again with no rotation (see Figure 11.12c).

Alternatively the plate motion could be accomodated by a wide zone of uniform deformation, where the rocks behaved viscously. The paleomagnetic vectors in the zone of deformation would all be rotated (Figure 11.12d). Such a situation does not happen in the rigid upper part of the crust, but it could happen at depth where the rocks might flow rather than break along fractures in response to tectonic forces.

One suggestion is that the rotated crustal blocks in southern California should be regarded as microplates. Such plates can be defined as segments of lithosphere (the rigid upper part of the earth's crust) that have been displaced with respect to adjacent plates along a complete set of boundary faults pene-

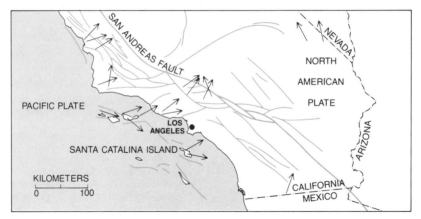

Figure 11.11 LARGE ROTATIONS ARE INFERRED for rocks in southern California on the basis of paleomagnetic declinations that deviate sharply from the craton declination, which is almost due north. The rocks range in age from 10 to 26 million years, a period in which the direction of the pole is known to have changed very little. The rotated rocks lie entirely to the west of the San Andreas fault on the Pacific plate, which is sliding northwestward parallel to the continental margin. Plate motion at a simple fault of this type, however, would not by itself be able to give rise to such large rotations, some of which amount to as much as five degrees per million years. These terranes must have been rotated by a more complex tectonic process.

trating all the way to the asthenosphere (the fluid region that begins about 100 kilometers below the lithosphere). One would expect the minimum length or width of a microplate to also be on the order of 100 kilometers. Inasmuch as many of the rotated blocks in southern California are much smaller than that, only 10 to 20 kilometers across, it appears that the faults that bound them are confined to the upper 15 kilometers of brittle upper crust and so do not penetrate the underlying ductile layer (Figures 11.12e,f). Rotational domains that small are better described as intracrustal blocks than as true microplates. Fault patterns of great complexity seem to be related to the evolving geology of the lower San Andreas region, including the geology involved in the origin and deformation of oil-bearing basins.

In western Oregon and Washington rotations ranging from 25 to 70 degrees have been found in rocks varying in age from 30 to 55 million years (see Figure 11.13). The largest rotations are found in the oldest rocks, which include lava flows and sediments that were originally emplaced on the ocean floor and are now accreted to the western edge of the continent to form Oregon's Coast Range. The Cascade Range, which lies to the east of the Coast Range and is younger, has been rotated clockwise about 25 degrees.

The tectonic environment in western Oregon is quite different from the one in Southern California. There are fewer earthquakes and the geologic formations have been less disrupted by faulting. J. Magill of Stanford and one of us (Cox) believe the rotations in western Oregon were produced in two phases by separate tectonic processes. The first phase of rotation was between 55 and 40 million years ago, when oceanic crust, which is the oldest part of the Coast Range, was being accreted to the continent. The second phase began about 20 million years ago and accompanied the well-documented thinning and stretching of the crust in the course of the extension of the Basin and Range province in eastern Oregon and Nevada. Whether the rotated blocks in western Oregon and Washington are true microplates or are shallow blocks of crust detached from the underlying lithosphere is an open question. The great length of the rotational domains in Oregon suggests that microplates may be involved, whereas in Washington the rotational domains are smaller, suggesting some decoupling from the rest of the lithosphere.

In speculating how the displaced terranes became accreted to North America we can begin with several useful observations. The first is that the leading edge of an accreted terrane does not, as one might expect, take the form of sutures typical of subduc-

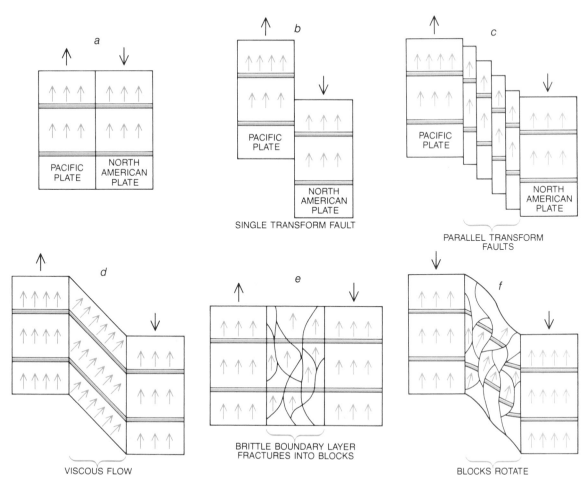

Figure 11.12 VARIETY OF DEFORMATIONS can result at a boundary where two plates slide past each other. Black arrows show relative motion; blue arrows show leomagnetic vectors embedded in the plates. Regardless of whether the relative motion of the two plates gives rise to a single fault (b) or to a series of faults parallel to the plate boundaries (c), the magnetic vectors remain unrotated. If the plates are separated by a zone in which rocks behave like a viscous fluid (d), the magnetic particles in the viscous rock will be rotated. If the crust between the two plates is brittle enough (e), it may crack. Motion of the plates can then cause blocks to rotate, carrying their paleomagnetic vectors (f).

tion zones. In a subduction zone the edge of an oceanic plate plunges at a steep angle under the continental margin. The terrane edges take the form of simple thrust faults or strike-slip faults. In a thrust fault one block merely moves up and over another block along a shallowly dipping fault. In a strike-slip fault two blocks slide past each other horizontally along a steeply dipping fault.

The second observation is that most terranes have become highly elongated and stretched parallel to

the edge of North America (see Figures 11.14 and 11.15). This is particularly true of the older terranes of Alaska and British Columbia, which on a small-scale geologic map look like thin strips of veneer applied to the edge of the continent.

That many of these terranes were transported to North America on oceanic plates can scarcely be questioned on the basis of the fossil and paleomagnetic evidence. This being so, the oceanic plate must have been consumed at a subduction zone when the terrane arrived at the continental margin. It is clear the terranes themselves survived the subduction process. The puzzling scarcity of subduction-zone sutures along present terrane boundaries implies that the sutures have been altered or hidden by subsequent geologic processes. Thrust faulting and strike-slip faulting are ubiquitous processes that are both capable of hiding the sutures.

A second puzzling feature of terranes is that many of them have survived the accretion process with little internal deformation. Since accretion implies collision at a subduction zone, one would ex-

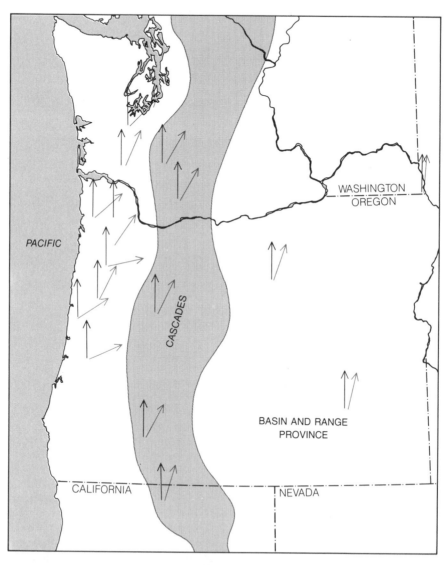

Figure 11.13 ROTATION OF ROCKS LESS THAN 60 MILLION YEARS OLD is observed in the terranes that form the western edge of Washington and Oregon. The black arrows show the direction of the paleomagnetic pole based on samples from the stable North American craton. (The arrows are here rotated slightly to the north for the sake of simplicity.) The colored arrows show the average paleomagnetic direction observed at each sampling site. All the rotations are clockwise. Moreover, the greatest rotations are found in the oldest rocks, which formed offshore on the ocean bottom and are now accreted to the continent west of the Cascade Range.

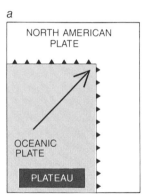
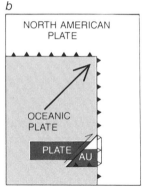
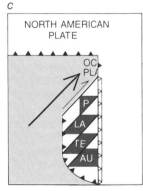

Figure 11.14 ELONGATION OF TERRANES may result if the subducting oceanic plate strikes the continental margin obliquely (*a*). As the plate plunges under the margin a plateau riding on the plate resists subduction and is accreted to the continent. When the plateau collides with North America (*b*), it splits off along a fault and is wedged into position as the rest of the plateau continues moving northeast. Process is repeated (*c*) as subduction zone (*barbed line*) jumps west.

pect the terranes to be much deformed. On the contrary, one finds large islands of relatively undeformed terranes such as Wrangellia in close proximity to more deformed and smaller terranes such as Chulitna. The extent to which a terrane is deformed in the course of accretion evidently depends on several factors: the velocity of the converging plates, the angle at which the plates collide, the width of the collision zone, the length of time the exotic terrane stays in the accretion zone and the strength of the terrane's rocks. In addition if a subduction zone gets clogged by a buoyant terrane, the zone may jump to a new position seaward of the newly accreted and largely undeformed terrane.

A description of three areas within the exotic terranes of western North America may illuminate the complexity and variety of the structural features evolved in the accretion process. In southwestern Alaska and adjacent British Columbia a complex suture zone involving Wrangellia and several other accreted terranes is well exposed in fjords that form deep bays along the coast. Evidently in mid-Cretaceous time, some 100 million years ago, Wrangellia collided with terranes that are now to the east. The collision resulted in intense deformation and metamorphism, followed by a major uplifting of the terranes to the east. The comparative recency of the collision is attested by the presence of deep-water, fine-grained sedimentary and volcanic rocks from late Jurassic to mid-Cretaceous time that were deposited in a deep marine basin landward of Wran-

gellia. Plutonic granites that intruded into the uplifted eastern terranes in the early Tertiary are largely undeformed, indicating that by then the accretion of Wrangellia to North America had been completed.

Farther north along the same boundary in southern Alaska an accretionary suture zone is beautifully preserved in the Alaska Range for several hundred kilometers east and west of Mount McKinley, but the geologic events recorded there are different from those in southwestern Alaska and British Columbia. In the Alaska Range, Jurassic and Cretaceous rocks that were deposited in a deep marine trough have been strongly deformed and telescoped to a small fraction of their original width and have been overridden by Wrangellia on the south along a major thrust fault. Scattered through the collapsed and disrupted basin are many fault-bounded small terranes, of which Chulitna is perhaps the most striking example. The origins and geologic histories of these small exotic terranes are totally unrelated to either Wrangellia or central Alaska, or for that matter to anything else known in North America. In the course of collision the small terranes were thrust over the younger strata of the deep marine trough, just as the Wrangellia terrane was. After the collision the entire region was further telescoped and deformed by dextral slip faulting, a process that continues to the present day.

The third area we shall describe lies in the Yukon

Territory farther to the east, where work by our Canadian colleagues suggests that the Stikine terrane first made contact with North America in the mid-Jurassic. This enormous terrane, probably the largest yet recognized, arrived on a plate bearing the roots of a volcanic arc and what appear to be oceanic materials of the Cache Creek terrane, which borders the Stikine terrane on the east. The collision eventually carried the oceanic and arc material up and eastward over the continental margin in the form of vast thrust sheets. Subsequent accretions added Wrangellia and other younger terranes to Stikine's trailing edge. This pileup of thrusted sheets along the ancient western edge of Norh America created a belt of new continental crust as much as 600 kilometers wide. Later folding and thrust faulting that extended into late Cretaceous and even early Tertiary time was accompanied by extensive strike-slip faulting that shifted large parts of the Canadian Cordillera (the entire complex of mountain ranges on the western side of the continent) hundreds of kilometers northward with respect to North America as a whole.

Accreted terranes play a major role in one of the most dramatic processes of global tectonics, the creation of mountain chains along convergent continental margins. Collisions between continental masses had been invoked long before the advent of plate tectonics to explain certain mountain belts, such as the Himalayas, that are deeply embedded within two large converging land masses. The possibility that collisions may also play a role in forming mountain belts that directly face an open ocean, such as the Andes of South America and the Cordillera of North America, has only recently been suggested. Here the collision is between the continent and much smaller land areas, including seamounts (isolated mountains on the sea floor), island arcs, marine plateaus and microcontinental blocks. The massive telescoping of crust, thrust faulting and metamorphism, however, are essentially similar to the consequences of collisions between continental masses. If one accepts that similar effects imply similar causes, it follows that massive, complexly deformed mountain systems imply collisions between separate, converging thick crustal blocks.

The idea of a collision between continents has been highly successful in explaining the massive telescoping of rock strata in the Himalayas, where the crust appears to have been shortened by 800 kilometers or more. Evidently the continental crust of India, some 40 kilometers thick, was too buoyant to be subducted very deeply at the suture zone where it collided with the Asian plate. Instead the converging crusts of India and Asia telescoped along thrust faults until the crust was twice the thickness of normal continental crust, forming the Himalayas. In the 40 million years since the initial collision the Indian subcontinent has continued to move northward, shoving Asian crustal rocks toward the north and east and causing massive disruption far into China. The continuing convergence is responsible for most of the devastating earthquakes that have racked the region.

The Andes are less well understood. Presumably they were created by the subduction of oceanic crust under continental crust, yet there is abundant evidence of compression and shortening in broad belts well inland from the subduction zone. The compression was attributed by one of us (Coney), and later by Kevin C. Burke of the State University of New York at Albany and Wilson, to a rapid movement of the continent toward the oceanic trench immediately above the subduction zone, where the continent meets the descending oceanic plate, creating a compressive stress that is transmitted inland from the continental margin.

An alternative model for mountain building in which terranes play a key role has recently been advanced by Zvi Ben-Avraham and Amos M. Nur of Stanford and two of us (Jones and Cox). In this model mountain building of the Andean type is more closely related to mountain building of the Himalayan collisional type than to simple subduction of oceanic crust. The model suggests that large oceanic plateaus, seamounts and volcanic ridges, some of which are comparable to continents in their thickness and density, may play the same role as the subcontinent of India in the creation of the Himalayas. Like India, these large masses of light rock are too buoyant to be subducted and so they serve to couple the forward motion of the lower subducting oceanic plate to the upper continental plate. On this view even the Andes may have been thrown up by the accretion of oceanic plateaus—perhaps exotic terranes not yet recognized—along the continental margin of South America. The difference in scale between the Andes and the Himalayas would reflect the difference between the width of the Indian subcontinent and the width of the plateaus responsible for the Andes.

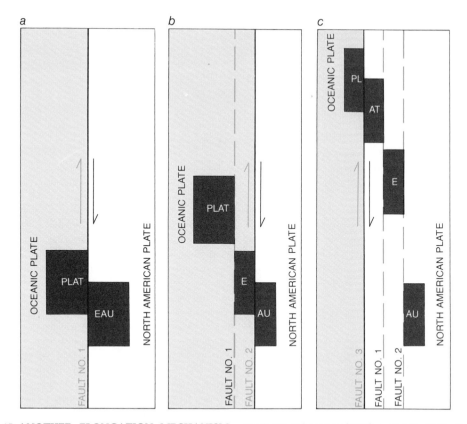

Figure 11.15 ANOTHER ELONGATION MECHANISM may involve the back-and-forth shifting of faults. After the plateau is lodged against the continental plate at the original docking line, Fault No. 1 develops to the west and begins carrying part of the plateau northward (*a*). In time Fault No. 1 becomes inactive, and Fault No. 2 develops to the east of it (*b*), slicing through the area of the plateau that had been landlocked. Still later Fault No. 2 becomes inactive and is succeeded by Fault No. 3, which further slices the plateau, carrying fragments of it northward.

A testable prediction of this model is that the mountain building should coincide with the accretion of the exotic terranes. An ideal test case is provided by the Laramide orogeny: the last great deformation of the North American Cordillera. The Laramide orogeny was an intense and widespread deformation and building of mountains between 40 and 80 million years ago. It took in a broad zone extending from the Sierra Nevada to the Rockies. It is one of the best described and least understood of all mountain-building episodes. Extending as far east as Denver, the Laramide orogeny produced the giant uplifts of the Rockies and the Colorado Plateau, which together give the Cordillera its extraordinary width. Most of the deformation of the Canadian Rockies and of the Sierra Madre Oriental of

eastern Mexico occurred at the same time. Throughout the period of deformation oceanic crust was being subducted along the west coast of North America. But how could subduction of ocean floor off California and Oregon be the cause of mountain building more than 1,200 kilometers to the east in Colorado?

Two alternative explanations based on plate-tectonic theory have been advanced by one of us (Coney). The first proposes that the angle at which the slab of oceanic lithosphere was subducted under North America was so shallow that even 1,500 kilometers from the coast the slab was still mechanically coupled to the overlying plate, thrusting it upward. The second explanation is that North America and the converging oceanic plate to the west were simply moving toward each other so fast that deformation took place at an unusually great distance from the subduction zone. Although there is some evidence supporting both explanations, many investigators have felt that even the two together are not adequate to account fully for such a wide and profound orogeny.

A third possibility is that the deformation was augmented by the arrival of terranes. Although most of the exotic terranes seem to have arrived before the Laramide orogeny, the event may represent the final phases of their collision with North America. The folding and faulting throughout the Cordillera would then represent the final "tightening up" of a poorly consolidated continental crust made up of newly accreted terranes. Interaction of the terranes with the adjoining ancient crust would presumably cause rotations, uplifts and overthrusts in a broad band encompassing both the North American craton and the terranes.

The driving force of such a process appears to have been the continuing subduction of the oceanic Pacific plates under North America rather than the successive arrival of new exotic terranes in the Cenozoic era. The collision of exotic terranes during the Laramide orogeny cannot, however, be totally ruled out. Recent paleomagnetic data obtained from central and southern California by David Howell,

Jack Vedder and Dwayne Champion of the U.S. Geological Survey have led to the suggestion that in Eocene time, no more than 50 million years ago, a large continental fragment derived from the latitude of southern Mexico collided with the southwestern margin of California.

The recognition of many exotic terranes in western North America adds an important new chapter to the geologic history of our continent. We propose that western North America has grown by more than 25 percent through accretion since early Jurassic time, a period of barely 200 million years. The growth was provided mainly by the addition of terranes that are of oceanic rather than continental origin. This implies real continental growth, not just the recycling of old continental material. Although the process of collision, accretion and continental growth is complex and poorly understood, there must have been much telescoping and transport of mass. The end result is new crust thickened by thrusting to continental proportions and added to the old continent. The concept of terranes being accreted piecemeal in western North America has important implications for the origin and evolution of the world's great mountain chains, many of which may have had a similar history.

The Supercontinent Cycle

Several times in earth history the continents have joined to form one body, which later broke apart. The process seems to be cyclic; it may shape geology and climate and thereby influence biological evolution.

. . .

R. Damian Nance, Thomas R. Worsley and Judith B. Moody
July, 1988

Is plate tectonics a random process or is it orderly? According to the theory of plate tectonics, the earth's rigid outer layer, called the lithosphere, is a mosaic of slablike plates that move with respect to one another at speeds averaging a few centimeters per year. The plates float on a hot, plastic layer of the earth's mantle called the asthenosphere. Most of the plate movements are driven by a process known as sea-floor spreading, in which molten material from the asthenosphere rises through the lithosphere at high ridges on the ocean floor, where it cools to become the crust that makes up the ocean bottom. Newly created oceanic crust moves steadily away from the midocean ridges toward the continents. If the sea floor and the adjacent continent are on the same lithospheric plate, the continent is carried along by the conveyor belt of oceanic crust. Alternatively, the oceanic crust may sink under the continent to rejoin the mantle, in a process known as subduction.

The continents are generally viewed as passive objects that are ferried about by sea-floor spreading. They are not entirely unchanged by the processes of plate tectonics, however. Separate blocks of continental crust can collide and merge, forming new, larger continents. Conversely, continents can be torn apart by deep rifts that eventually become the centers of new ocean basins. Indeed, there is evidence that several times in the history of the earth the continents have undergone these processes on a grand scale: several times most or all of the continents have gathered to form a single supercontinent, which has later split into many smaller continents only to rejoin and form a supercontinent again.

What governs the formation and destruction of supercontinents? Do they appear and disappear simply by chance, because of the random shifting of continental plates? Various regularities in the geologic record have led the three of us to believe that a much more orderly, even cyclic, process must be at work. Drawing on the ideas of Don L. Anderson of the California Institute of Technology and on the prescient observations of the Dutch geologist J. Umgrove (set out in his 1947 book *The Pulse of the Earth*), we have devised a theoretical framework that describes what may be the underlying mechanisms of such a "supercontinent cycle."

In our theory the dominant force comes from heat. It is generally understood that tectonic plates are driven by convective motions in the underlying mantle, which are powered by heat from the decay of radioactive elements. The radioactive decay (and the resulting production of heat) is a continuous process whose rate has declined smoothly with time, and so the production of heat cannot in itself account for the episodicity inherent in an alternation between continental assembly and continental breakup.

The key phenomenon, we think, is not the production of heat but rather its conduction and loss through the earth's crust. Continental crust is only half as efficient as oceanic crust at conducting heat. Consequently, as Anderson has pointed out, if a stationary supercontinent covers some part of the earth's surface, heat from the mantle should accumulate under the supercontinent, causing it to dome upward and eventually break apart. As fragments of the supercontinent disperse, heat can be transferred through the new ocean basins created between them. After a certain amount of heat has escaped, the continental fragments may be driven back together.

In other words, we think the surface of the earth is like a coffee percolator. As in a coffee percolator, the input of heat is essentially continuous. Because of poor conduction through the continents, however, the heat is released in relatively sudden bursts.

This theoretical frameork and its corollaries make it possible to tie together a number of observations in widely disparate fields. They make it possible, for example, to understand the timing of the extreme changes in sea level that have taken place in the past 570 million years. The framework also helps to explain and link many other events of the past 2,500 million years, such as periods of intense mountain building, episodes of glaciation and changes in the nature of life on the earth. The supercontinent cycle, in our view, is a major driving process that has provided the impetus for many of the most important developments in the earth's history.

THE OPENING OF OCEANS

Our model builds on an earlier description of episodic plate motions known as the Wilson cycle. Named for J. Tuzo Wilson of the Ontario Science Center, the Wilson cycle is the process by which continents rift to form ocean basins and the ocean basins later close to reassemble the continents. In the first stage of the Wilson cycle volcanic "hot spots" form in a continent's interior; the hot spots are then connected by rift valleys, along which the continent eventually splits. When the continent fragments, the rift valleys grow to become a new ocean as hot mantle material wells up through the rifts to form the sea floor. The continental fragments move apart, sliding away from these elevated "spreading centers" as mantle material wells up.

As the material making up the ocean floor ages, it cools, becomes denser and subsides, increasing the depth of the ocean. Eventually, about 200 million years after the first rift formed, the oldest part of the new ocean floor (the part directly adjacent to the continental fragments) becomes so dense that it sinks under the continental crust: it is subducted. The processes of subduction then close the ocean, bringing the continents back together. Eventually the continents collide and rejoin, and the compressive forces of collision create mountain belts.

When viewed in terms of Wilson cycles, there is a striking contrast between the evolution of the continental margins surrounding the North Atlantic and the margins of the Pacific. The margins of the North Atlantic have undergone a series of Wilson cycles during the past billion years; the regions bordering the Pacific have apparently undergone none. In other words, oceans have repeatedly opened and closed in the vicinity of the present-day North Atlantic, while a single ocean has been maintained continuously in the vicinity of the Pacific.

In our model, then, the Pacific is the descendant of the oceanic hemisphere that has surrounded each incarnation of the supercontinent; each of the Wilson cycles that took place in what is now the North Atlantic region occurred as part of the breakup and reassembly of a supercontinent. The Atlantic should therefore be expected to close again, once more reuniting the continents in a supercontinent surrounded by a single superocean.

At present the sea-floor crust of the Pacific is being subducted under all the continents that surround it, whereas the floor of the Atlantic generally butts up against surrounding continental blocks. In our framework this means that the continents are still in the process of dispersing after the breakup about 200 million years ago of the most recent supercontinent, which Alfred Wegener, the father of the theory of continental drift, christened Pangaea (see Figures 7.4 and 12.1), or "all earth" [see "The Breakup of Pangaea," by Robert S. Dietz and John C. Holden; SCIENTIFIC AMERICAN, October, 1970]. This supercontinent apparently formed some 300 million years ago by assembly of formerly separate continents. One result of that assembly was the formation of the Appalachian mountains by the collision of proto–North America with proto-Africa, (see Figure 10.6). Other orogenic belts that formed during the final assembly of Pangaea include the Urals and the Mauritanides (see Figure 10.8).

About 200 million years ago, heat accumulating under the supercontinent broke through in rifts that eventually became oceans. The growth of these shallow oceans at the expense of the older and deeper superocean, Panthalassa (see Figure 7.4), raised sea level, partially drowning the continents. Sea level rose to a maximum about 80 million years ago then fell, as the new oceans became older and deeper and the world's present geography was established. The continents are now approaching their maximum dispersal. Soon (on a geologic time scale) the crust of the Atlantic will become old and dense enough to sink under the surrounding continents, beginning the process that will close the Atlantic ocean basin.

SURPRISING REGULARITY

A second underpinning of our supercontinent-cycle hypothesis is the timing of various episodes of mountain building and episodes or rifting. The ages of mountain ranges that could have been produced by the compressive forces that accompany continental collisions reveal a surprising regularity. This kind of mountain building was particularly intense, occurring in several parts of the world, during six distinct periods. The periods were broadly centered on dates about 2,600 million years ago, 2,100 million years ago, a time between 1,800 and 1,600 million years ago, 1,100 million years ago, 650 million years ago and 250 million years ago (see Figure 1.2). The timing shows a certain periodicity: the interval between any two of these periods of intense compressive mountain building was about 400 to 500 million years.

What is more, about 100 million years after each of these periods of mountain building there appears to have been a period of rifting (see Figures 1.2 and 1.4). Large numbers of mantle-derived rocks — rocks that may have been produced when magma welled up into cracks created by rifting — date from times broadly centered on 2,500 million years ago, 2,000 million years ago, a time between 1,700 and 1,500 million years ago, 1,000 million years ago and 600 million years ago. The mountain building of

250 million years ago, of course, was followed by the rifting and eventual breakup of Pangaea.

These regularities indicate to us that supercontinents are created in a cyclic process, in which one complete cycle takes about 500 million years (see Figure 12.2). By examining these geologic records and others, and by taking into account such factors as the rate at which seafloor spreading takes place in present-day oceans, we have calculated a more precise timetable for the supercontinent cycle. After the fragments of a supercontinent first separate — probably some 40 million years or so after rifting begins — we estimate that it should take about 160 million years for the fragments to reach their greatest dispersal and for subduction to begin in the new oceans. After the continents begin to move back together, another 160 million years or so should elapse before they re-form a supercontinent. The supercontinent should survive for about 80 million years before enough heat accumulates under it to cause rifts to form. Forty million years later that rifting will lead to another breakup, 440 million years after the previous one.

EFFECTS ON SEA LEVEL

How can one test whether the supercontinent cycle proceeds as we have described? The cycle is likely to have striking effects on sea level, for which there are clear geologic records covering the past 570 million years. Assuming a constant amount of water in the world oceans, sea level (in relation to continental mass) is largely determined by two factors: the total volume of the world's ocean basins, which depends in part on the average depth of the sea floor, and the relative elevation of continents. The supercontinent cycle would involve the creation and destruction of ocean basins and the thermal uplifting of continents, and so it should have a profound influence on both factors (see Figure 12.3).

As the material making up the ocean floor moves away from midocean ridges during sea-floor spreading, it cools and subsides, its depth increasing as the square root of its age. Wolfgang H. Berger and Edward L. Winterer of the Scripps Institution of Oceanography have calculated how the average age of the world ocean floor should change during the breakup of a supercontinent. Before the breakup the average age of the ocean should remain constant, because in the superocean surrounding the supercontinent new sea floor is created at about the same rate as old sea floor is destroyed by subduction

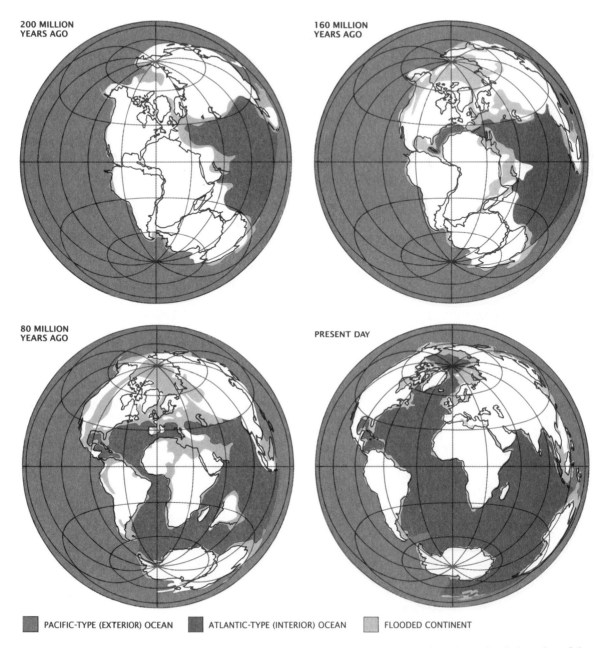

PACIFIC-TYPE (EXTERIOR) OCEAN	ATLANTIC-TYPE (INTERIOR) OCEAN	FLOODED CONTINENT

Figure 12.1 WHOLE EARTH MAPS illustrating the breakup of Pangaea. The upper left diagram shows the configuration of land about 200 million years ago; the dark blue ocean on the east side of the supercontinent is the Tethys (Figure 11.3). The progressive configurations at 160 million years ago, 80 million years ago and present day illustrate the dispersal of continents by the opening of the Atlantic and Indian oceans, and the resultant closure of the Tethys Ocean to form the Alpine-Himalayan mountains (see Figures 1.2, 9.1 and 9.4). (Maps based on work by A. G. Smith and J. C. Briden.)

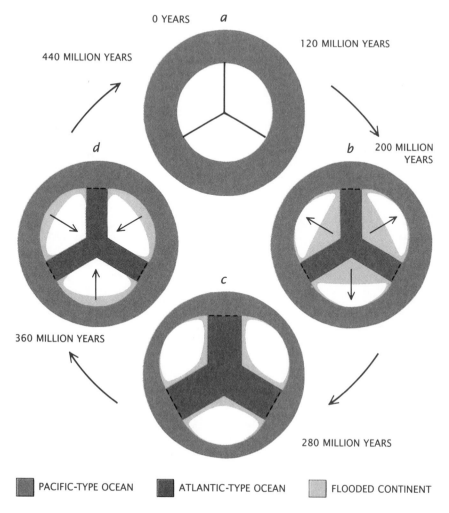

Figure 12.2 SUPERCONTINENT CYCLE is depicted schematically. A supercontinent (*a*) can survive for about 80 million years before the accumulation of heat causes rifts to form, and for another 40 million years or so before it is torn apart into separate continents. The continents drift apart (*b*) until they reach their greatest degree of dispersal, about 160 million years later (*c*). Then they move back together (*d*), eventually reforming the supercontinent (*a*). The entire cycle takes about 440 million years.

under the landmass. During the breakup, the subducting, "Pacific type" superocean will be replaced by an increasing proportion of nonsubducting, "Atlantic type" oceans. Later, when the supercontinent begins to reunite, these "interior" oceans will be destroyed by subduction and replaced by Pacific-type ocean again. These processes affect the average age of the world ocean floor.

Immediately after the breakup of a supercontinent the world ocean floor should, on the average, become progressively younger and shallower as young, Atlantic-type oceans begin to replace the older, Pacific-type ocean. When the Atlantic-type oceans reach the same average age as the Pacific-type ocean, the trend should reverse: the growth of increasingly old Atlantic-type oceans should cause the world ocean floor to age and deepen. The maximum average depth should occur when Atlantic-type oceans reach their greatest average age, just before they begin to be subducted. Then, as the

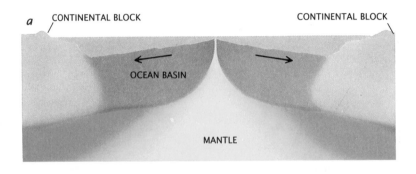

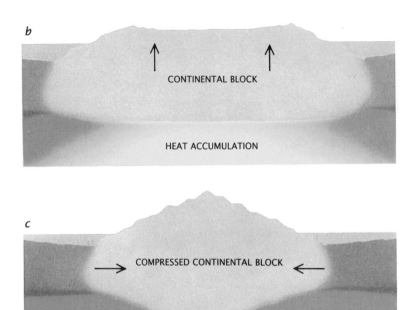

Figure 12.3 SEA LEVEL with respect to the continents is controlled by several tectonic factors. One is the age of the sea floor, which is created by the upwelling of hot material from the earth's mantle at midocean "spreading centers" (*a*). As the sea floor spreads it cools, becomes denser and sinks; older ocean is therefore deeper, and so sea level becomes lower when the average age of the world's oceans increases. The accumulation of heat under stationary continental crust (*b*) alters sea level by buoying the continent upward. Sea level is also affected by compression (*c*) or extension (not shown) of continents. When continents are compressed, the total area of the world ocean increases while the volume of water remains constant: sea level is lowered.

oldest areas of the Atlantic-type oceans are subducted and the oceans close, the world ocean floor should become younger and shallower again.

Calculations of sea level based on these parameters alone suggest that a supercontinent's continental shelves should be flooded, because the ocean basin surrounding a supercontinent is younger and shallower than, for example, the floor of today's world ocean. A second factor, however, must be added to the analysis: the degree to which a supercontinent would be uplifted by the heat that would accumulate under it. If the supercontinent is lifted high enough, sea level in relation to the continental mass could still be low even if the sea floor is comparatively shallow (see Figure 12.4).

One way to estimate how much a supercontinent might be uplifted thermally is to consider present-day Africa. Africa has remained essentially stationary for at least the past 200 million years, during which time a good deal of heat from the mantle has accumulated under it. (Some of that heat is being released in the rift valleys now forming in various areas of the continent.) By comparing the height relative to sea level of Africa's shelf break (the true edge of the continent) with the height of the shelf breaks of other continents, we can estimate that thermal uplifting has buoyed Africa by about 400 meters. As a lower limit, then, one can expect that a supercontinent would be thermally uplifted by at least 400 meters.

Other factors should also cause the supercontinent to be emergent (elevated in relation to sea level). For example, the collisions that take place during the assembly of the supercontinent should compress and thicken continental crust, decreasing the earth's total land area. This would add to the total area of the ocean basins and thereby lower sea level. Conversely, the stretching and extension of crust that accompanies the breakup of a supercontinent should lower the total area of the world ocean basin, thereby raising sea level (see Figure 12.5).

By adding together all these components, it is possible to determine how sea level should change in the course of every phase of the supercontinent cycle. As we have noted, during the existence of a supercontinent sea level should be relatively low. As the supercontinent breaks up, sea level should rise, both because the continental fragments will stretch and subside thermally and because the

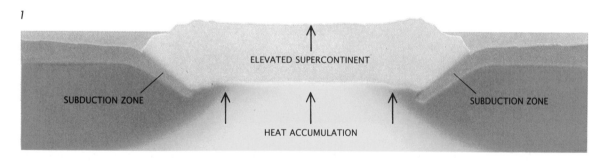

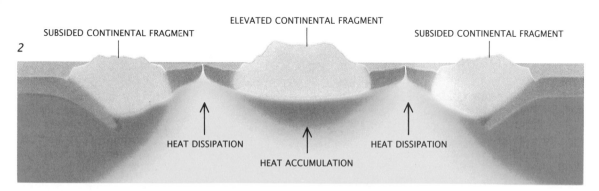

Figure 12.4 HEAT controls the elevation of a supercontinent and its fragments. A supercontinent (1), surrounded by subduction zones (where sea-floor material sinks under the continent), remains stationary in relation to the underlying mantle. Heat accumulates under it, buoying it upward. After the supercontinent breaks up (2), fragments subside as they drift away. A fragment that stays in place (*center*) remains elevated. Present-day Africa is one example of such a stationary fragment.

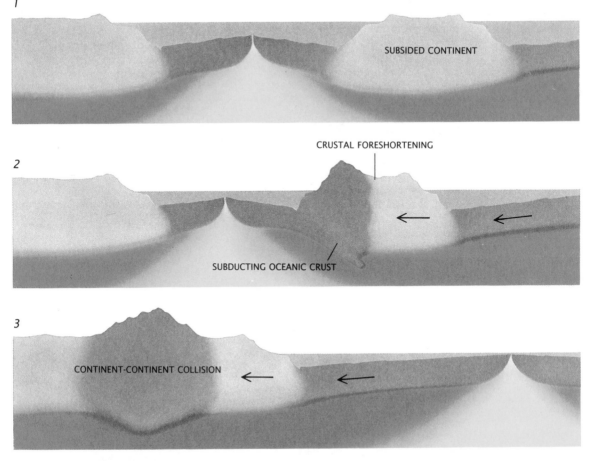

Figure 12.5 FORESHORTENING of continental blocks increases the total area of the world ocean and thereby lowers the global sea level. Just after the breakup of a supercontinent (*1*) the new sea floor (*left center*) butts up against continental fragments. Later, as the new ocean basin closes (*2*), the continent moves over the sea floor, subducting it; the resulting compressive forces foreshorten the continent and raise mountains, in the process lowering sea level. Still later, when continents collide during reassembly of the supercontinent (*3*), the continental crust is foreshortened even further, again lowering sea level.

breakup will replace old, Pacific-type ocean with young, Atlantic-type ocean. Sea level should continue to rise for about 80 million years, as the younger oceans make up a greater fraction of the world ocean. Then, as the Atlantic-type oceans age and expand, sea-level should decline for another 80 million years or so, until the Atlantic-type oceans begin to be subducted.

When the continents begin to come together, sea level should rise, as older Atlantic-type crust is subducted. That rise in sea level should continue for another 80 million years, until the supercontinent begins to be reassembled. Then, as continents collide and the growing supercontinent is uplifted thermally, sea level should decline for about 80 million years. Once the supercontinent has been formed, sea level should remain static for another 120 million years, until the supercontinent breaks up again.

These predictions of changes in sea level match the geologic record of the past 570 million years, which is as far back as sea level can be determined with any reliability. In particular, the timing and the relative magnitudes of sea-level changes predicted by the model match the preserved record. The absolute values, of course, always vary from continent to continent; the model gives only global averages.

TESTING THE MODEL

Evidence for the supercontinent cycle can also be found by examining the isotopes of sulfur and carbon found in certain marine sediments. (Isotopes are atoms of the same element that have different atomic weights.)

During the early stages of breakup, a supercontinent is likely to include a number of marine rifts that, like the modern Red Sea, are weakly connected to the world ocean. These rifts can undergo a continuous process of partial evaporation, in which certain elements, such as sulfur, precipitate out of the seawater to form minerals. When seawater containing sulfur is evaporated, heavy sulfur (which has an atomic weight of 34) precipitates out more readily than light sulfur (which has an atomic weight of 32).

If an evaporating marine rift continues to mix with the world ocean, it should act as a sink for heavy sulfur: it should tend to pull heavy sulfur from the ocean as a whole and bury it in evaporitic sediments. Then the world ocean should become relatively depleted in heavy sulfur and enriched in light sulfur. Hence in sediments derived from the world ocean as a whole during the period of the supercontinent, we should expect to see relatively high levels of light sulfur and low levels of heavy sulfur. This is indeed what is found in open marine sediments formed about 200 and 600 million years ago—that is, during the past two instances of the supercontinent.

Carbon isotopes also give evidence of supercontinents. The lighter isotope of carbon (carbon 12) diffuses in solution faster than a heavier isotope (carbon 13). As a result light carbon is more likely to be taken up by organisms and incorporated into their biomass. Organisms are therefore a sink for light carbon. During periods of low sea level the rate of organic productivity in the world ocean should be high, because greater amounts of nutrients such as phosphorus and nitrogen—eroded from continental crust and carried to the sea by rivers—will be available when more continental crust is exposed.

Thus when sea level is low, more carbon (particu-

larly light carbon) will be incorporated into organisms, and the water of the world ocean will be comparatively depleted in light carbon and enriched in heavy carbon. When examining such seawater-derived sediments as limestone (calcium carbonate), then, we should expect to find relatively high levels of heavy carbon and low levels of light carbon if the sediments were produced during a period of low sea level. In similar sediments produced during a period of high sea level, we should expect to find relatively more light carbon and less heavy carbon. And indeed, the ratio of heavy carbon to light carbon in such sediments closely matches the predictions of our model for the past 600 million years.

CLIMATE AND LIFE

Perhaps the most important effects of the supercontinent cycle are its influences on climate and life. What should those effects be? Most of the climatological effects of the supercontinent cycle will be driven by the changes in sea level that are caused by the processes of continental breakup, dispersal and reassembly.

When sea level is low—that is, when the world is dominated by a single emergent supercontinent or when individual continents are widely dispersed (as they are now) and the world ocean floor is at its oldest—large amounts of silicate minerals in the continental crust, such as calcium silicates, are exposed to weathering and erosion; they are dissolved in rivers and carried into the world ocean. When these dissolved silicates are mixed into seawater, they combine with dissolved carbon dioxide to produce solid precipitates [see "How Climate Evolved on the Terrestrial Planets," by James F. Kasting, Owen B. Toon and James B. Pollack; SCIENTIFIC AMERICAN, February, 1988]. For example, calcium silicates may combine with carbon dioxide to produce calcite (limestone) and quartz. This process draws down carbon dioxide from the atmosphere.

Carbon dioxide in the atmosphere helps the earth to retain the heat it gains from solar radiation. When carbon dioxide is drawn down into oceanic deposits, this "greenhouse" warming effect is diminished, and the world climate becomes colder. If there is an emergent continental landmass near enough to the pole, glaciers will form (as they have on modern-day Antarctica and Greenland).

Glaciation has several important effects on the global climate. For one thing, it removes water from the ocean basins, causing sea level to drop still lower. Glaciation also amplifies circulation and mixing in the world ocean. Much of the ocean circulation of the present-day earth, for example, is driven by a global "heat engine," in which warm, salty water from the Tropics and the subtropics flows toward the pole, where it gives up its heat, sinks to the bottom and flows back toward the Equator [see "Polynyas in the Southern Ocean," by Arnold L. Gordon and Josefino C. Comiso; SCIENTIFIC AMERICAN, June, 1988]. By mixing surface water into the deep water, the heat engine distributes oxygen and other nutrients throughout the ocean. Ice at the pole keeps the polar waters cold, helping to maintain the temperature difference that drives the heat engine.

Such vertical circulation combines with the increased supply of continent-derived nutrients to raise the level of biological productivity at times of low sea level. High productivity should trap still more carbon in organic matter, further lowering levels of atmospheric carbon dioxide.

The environments that are most hospitable to marine life are the continental shelves, where continent-derived nutrients are abundant and shallow depth allow sunlight to penetrate to the sea floor. When biological productivity is high, however, these shelf environments are unavailable, exposed by the low sea level. As a result many established species will become extinct, and new, innovative species will be favored. This does not mean, however, that life will be particularly diverse. On the contrary, such conditions—high nutrient levels but few available environmental niches—lead to ecosystems in which a large amount of biomass is concentrated in relatively few successful species.

CLIMATE AFTER BREAKUP

Thus the effects of low sea level include a propensity toward glaciation, strong vertical circulation in the world ocean, high biological productivity, biological innovation and a low degree of biological diversity. What should be the effects of high sea level, which would be expected just after a supercontinent breaks up or just before it is reassembled?

When continents are drowned, a relatively small amount of silicates will be available to sequester atmospheric carbon dioxide in sea-floor deposits. Meanwhile carbon dioxide will be released into the ocean and from there into the atmosphere by the hot mantle material that wells up at sea-floor spreading centers. Also, the subduction of oceanic crust and the resulting melting of limestone deposits in the subducted sediments will release still more carbon dioxide into the atmosphere at the volcanoes that mark subduction zones.

Atmospheric levels of carbon dioxide will therefore rise and the earth's climate will become warmer. Polar ice caps should melt, raising sea level still higher and further drowning the continents. The absence of polar ice will reduce vertical and horizontal circulation of the world ocean, causing it to begin to stagnate: oxygen and nutrient levels in ocean waters will decline, and with them biological productivity. On the other hand, the drowned continental crust will provide a large area for the shallow seas that are most hospitable to life. The resulting ecosystems will resemble those of the present-day Tropics, where the climate is warm, nutrient levels are low and a relatively large number of environmental niches are available. Like today's Tropics, these ecosystems would be characterized by low productivity and great species diversity.

How well does the record of past climates and life forms bear out these predictions? One of the most impressive confirmations of our model is found in records of glaciation. All the known episodes of glaciation in the history of the earth took place at times when according to our model sea levels should have been low. The converse is not true. That is, not every period when sea levels should have been low had an episode of glaciation; probability dictates that in some periods there would not have been an emergent continent near the pole.

BIOLOGICAL EVIDENCE

The biological record is a little more ambiguous, for a variety of reasons. Perhaps most important, the fossil record is not uniform throughout time. The record is based mainly on deposits buried on continental crust—buried when sea levels were high. When sea levels are low, marine organisms will generally live offshore, beyond the exposed continental shelves. Deposits recording these periods are rare: such deposits are likely to have been destroyed later, by subduction of the ocean floor. Nevertheless, the available evidence tends to confirm our hypothesis.

For example, the few geologic records of marine life during the existence of the most recent supercontinent, Pangaea, indicate low diversity of species, as we would expect when sea levels are low. In addition, the period of drowning that followed the breakup of Pangaea was characterized by high levels of species diversity.

Looking further back in time, the breakup of the previous supercontinent about 600 million years ago was marked by the first recorded appearance of shelled animals. During the period following this breakup there was what has been called an explosion of diversity [see "The Emergence of Animals," by Mark A. S. McMenamin; SCIENTIFIC AMERICAN, April, 1987]. In particular, shelled animals radiated into a highly diverse array.

Looking back still further, the first recorded multicellular animals are found in marine sediments that are about a billion years old. These sediments would have been deposited, according to our model, right after the breakup of a supercontinent. It is quite possible that this biological innovation occurred during the existence of the supercontinent but was not recorded until the continent broke up and sea levels rose, drowning the continental shelves.

A still older innovation may also be linked to the supercontinent cycle. About 2,100 million years ago, just prior to an assumed supercontinental breakup, blue-green algae first developed heterocysts, the organelles that make it possible to fix nitrogen (to crack apart nitrogen molecules and bind the constituent atoms to carbon in organic matter) even in the presence of atmospheric oxygen. Without heterocysts or similar organelles, the chemical reactions of nitrogen fixation can be interrupted by oxygen atoms that bind to the nitrogen atoms. The atmosphere was then just beginning to contain oxygen; the innovation made it possible for many later organisms—the predecessors of today's photosynthetic plants—to survive in the new, oxygen-bearing atmosphere, which otherwise could have been poisonous to them.

A NEW FRAMEWORK

The supercontinent-cycle hypothesis represents a new framework, a new way to understand the geologic history of the earth. It suggests that the processes of plate tectonics on the largest scale are primarily governed not by chance but by a regular, cyclic process.

The supercontinent cycle also represents a new way of understanding the history of life on the earth. The large-scale climatological effects brought about by various phases of the supercontinent cycle —continental drowning or emergence, glaciation and ocean circulation, stagnation in the world ocean and other effects—drove many of the important biological innovations that have directed the later course of evolution. In a sense, then, the supercontinent cycle is indeed the pulse of the earth: with every beat the earth's climate, geology and population of living organisms are advanced and renewed.

The Authors

ELDRIDGE MOORES is professor of geology at the University of California at Davis. He earned his B.S. degree in geology from the California Institute of Technology in 1959 and his M.A. (1961) and Ph.D. (1963) in geology from Princeton University. From 1963 to 1966 he was a postdoctoral fellow at Princeton studying the Vourinos ophiolite complex in northern Greece, then joined the faculty of the University of California. From 1981 to 1988 he edited the journal *Geology*. He has worked extensively in the eastern Mediterranean, principally in Greece and Cyprus, and in western North America, primarily in the Sierra Nevada of California and surrounding regions.

B. CLARK BURCHFIEL ("The Continental Crust") is professor of geology at the Massachusetts Institute of Technology. He earned his B.S. in 1957, and his M.S. in 1958, both from Stanford University. He received his Ph.D. in geology from Yale University in 1961. In the same year he joined the faculty at Rice University; he was at Rice until 1976, when he moved to M.I.T. Burchfiel writes: "My current interest is in the process of orogenesis (mountain building) and its relation to the interactions of tectonic plates and to intraplate deformation. The specific areas I study are western North America, the Alpine mountain chains of eastern Europe and (more recently) of north-central China." He is a member of the National Academy of Sciences.

JEAN FRANCHETEAU ("The Oceanic Crust") is a physicist on the staff of the Institut de Physique du Globe of the University of Paris. He is a native of France who received his diploma in mining engineering at the École des Mines in Nancy. He came to the U.S. to continue his education; the Scripps Institution of Oceanography awarded him a Ph.D. in geophysics in 1970. He returned to France to join the staff of the Centre Océanologique de Bretagne in Brest. He left the center in 1981 to go to the University of Paris. Francheteau's main scientific interest is the exploration of the ocean floor. His work has concerned the structure and morphology of the oceanic crust and the tectonics and geophysics of the ocean bottom. He was the leader of Project RITA in 1978.

GREGORY E. VINK, W. JASON MORGAN and **PETER R. VOGT** ("The Earth's Hot Spots") specialize in marine geophysics. Vink is a staff geophysicist with the Institute for Research in Seismology in Washington, D.C. He received his B.A. at Colgate University in 1979 and his Ph.D. from Princeton University in 1983; his dissertation described the tectonic evolution of the Norwegian-Greenland Sea and the Arctic Ocean. Morgan is professor of geophysics at Princeton and a member of the National Academy of Sciences. He went to the Georgia Institute of Technology as an undergraduate and received a Ph.D. in physics from Princeton in 1964, where he has remained. Vogt is a staff geophysicist at the Naval Research Laboratory. He was educated at the California Institute of Technology and the University of Wisconsin, which granted him a Ph.D. in oceanography in 1968. He worked in the U.S. Naval Oceanographic Office until 1975 and then joined the Naval Research Laboratory; he holds a concurrent post at the University of Oslo.

DON L. ANDERSON and **ADAM M. DZIEWONSKI** ("Seismic Tomography") are geophysicists who have taught together at the California Institute of Technology. Anderson is a professor there and director of Caltech's Seismological Laboratory. He earned his B.S. from the Rensselaer Polytechnic Institute in 1955, then worked as an exploration seismologist for the Chevron Oil Company. In 1956 he joined an Air Force geophysical research team and led six expeditions to Greenland to study the elastic properties of sea ice. He went to Caltech for his master's (1959) and his doctorate (1962) and subsequently joined the faculty. He is a member of the National Academy of Sciences. Dziewonski is chairman of the department of geological sciences at Harvard University and spent an academic year at Caltech as a Fairchild Distinguished Scholar. He was educated in Poland, earning an M.S. at the University of Warsaw in 1960 and a Doctor of Technical Sciences degree from the Academy of Mines and Metallurgy in Cracow in 1965. He came to the U.S. in 1965 as a postdoctoral fellow at the Southwest Center for Advanced Studies (now the University of Texas at Dallas). In 1969 he became a member of the faculty there, and in 1972 he accepted a position at Harvard.

VINCENT COURTILLOT and **GREGORY E. VINK** ("How Continents Break Up") are geophysicists. Courtillot is professor of geophysics and chairman of the department of earth sciences at the University of Paris. He attended the Paris School of Mines as an undergraduate. He came to the U.S. to pursue his studies, earning his M.S. in geophysics at Stanford University in 1972 before returning to France; his Ph.D. in geophysics was awarded in 1977 by the University of Paris. Vink is a staff geophysicist with the Institute for Research in Seismology in Washington, D.C. He received his B.A. at Colgate University in 1979 and his Ph.D. from Princeton University in 1983; his dissertation described the tectonic evolution of the Norwegian-Greenland Sea and the Arctic Ocean.

ENRICO BONATTI and **KATHLEEN CRANE** ("Oceanic Fracture Zones") are members of the staff of the Lamont-Doherty Geological Observatory of Columbia University. Bonatti, a native of Italy and a senior research scientist at the observatory, obtained degrees in geological science at the University of Pisa and the Scuola Normale Superiore in Pisa. He came to the U.S. on a Fulbright fellowship in 1960, spending one year at Yale University and four years at the Scripps Institution of Oceanography. He was professor of marine geology at the University of Miami for several years before moving to Lamont in 1976. Crane holds a B.A. from Oregon State University and a Ph.D. from the Scripps Institution of Oceanography. She writes: "During my years in oceanography and many expeditions at sea I became interested in the vertical dynamics of ocean crust. I joined forces with Enrico Bonatti in the Red Sea, where a newly forming midocean ridge is ramming into the Egyptian coast, thereby pushing up the island of Zabargad."

DAVID G. HOWELL ("Terranes") is a research geologist for the U.S. Geological Survey. He was graduated from Colgate University in 1962 and did graduate studies at the University of California at Santa Barbara, where he received a master's degree in 1969. He served three years in the U.S. Army before earning his Ph.D. from Santa Barbara in 1974. He then joined the Geological Survey. In 1980 he took a concurrent post as consulting professor at Stanford University.

IAN G. GASS is professor of earth sciences at the Open University in England. Born in England, he spent his early childhood in Burma. After four years of military service during World War II he returned to the University of Leeds to study geology, receiving his B.Sc. in 1952. He served in the geological surveys of the Sudan and Cyprus, then involuntarily retired when the country achieved independence. He obtained his Ph.D. from Leeds in 1960. After teaching briefly at the University of Leicester, he joined the faculty at Leeds and remained there until 1969, when he moved to the Open University.

PETER MOLNAR ("The Structure of Mountain Ranges") is senior research associate of earth, atmospheric and planetary sciences at the Massachusetts Institute of Technology, a position he has held since 1978. He is a graduate of Oberlin College and received his Ph.D. in geophysics from Columbia University in 1970. Molnar writes: "I try to spend at least three months each year in remote mountainous environments doing fieldwork and savoring the virtues of the particular indigenous culture." He recently was elected (1990) to membership in the National Academy of Sciences.

FREDERICK A. COOK, LARRY D. BROWN and **JACK E. OLIVER** ("The Southern Appalachians and the Growth of Continents") are geophysicists. Cook graduated from the University of Wyoming in 1973. He completed his master's degree there two years later and worked for the Continental Oil Company before moving to Cornell, where he received his Ph.D. He is a professor of geophysics at the University of Calgary, Canada. Brown has been a faculty member of geological sciences at Cornell since 1977. He did undergraduate work in physics at the Georgia Institute of Technology and received his Ph.D. in geophysics from Cornell in 1976. Oliver has served since 1971 as Irving Porter Church Professor of Engineering and as chairman of the department of geological sciences. He obtained a B.A. from Columbia College in 1947 and a master's degree in physics and Ph.D. in geophysics from Columbia University in 1950 and 1953. From 1955 through 1971 he was on the faculty at Columbia, spending the last three of those years as chairman of its department of geology. He is a member of the National Academy of Sciences.

DAVID L. JONES, ALLAN COX, PETER CONEY and **MYRL BECK** ("The Growth of Western North America") are geophysicists with a common interest in microplate tectonics. Jones is geologist with the Western region of the U.S. Geological Survey. His B.S. (1952) is from Yale University; his M.S. (1953) and Ph.D. (1956) are from Stanford University. Cox was dean of the School of Earth Science at Stanford prior to his death in 1987. His bachelor's degree and doctorate in geophysics were from the University of California at

Berkeley. He worked for the U.S. Geological Survey or for Stanford, sometimes in combination. He was a member of the National Academy of Sciences. Coney is professor of geosciences at the University of Arizona. His B.A. (1951) is from Colby College; his Ph.D. in geology (1964) is from the University of New Mexico. Beck is professor of geology at Western Washington University. His bachelor's degree and master's degree in geology are from Stanford. After receiving his master's degree he worked for the Standard Oil Company of California and the U.S. Geological Survey before returning to get his Ph.D. at the University of California at Riverside in 1969; he went to Western Washington in the same year.

R. DAMIAN NANCE, THOMAS R. WORSLEY and **JUDITH B. MOODY** ("The Supercontinent Cycle") have combined their respective specialties of tectonics, oceanography and geochemistry in a particularly close partnership. Nance teaches at Ohio University. He received his Ph.D. in 1978 from the University of Cambridge, and he taught at St. Francis Xavier University in Nova Scotia before going to Ohio in 1980. Moody is president of J. B. Moody and Associates; Moody and Worsley have been married for nine years. Worsley also teaches at Ohio. He received his Ph.D. in 1979 from the University of Illinois; he went to Ohio in 1977 after teaching at the University of Washington. Moody is president of J. B. Moody and Associates in Columbus, Ohio. She received her Ph.D. (1974) from McGill University. From 1981 until 1988 she worked at the Battelle Memorial Institute.

Bibliographies

1. The Continental Crust

Wilson, J. T., ed. 1976. *Continents adrift and continents aground.* W. H. Freeman and Company.

Tarling, D. H., ed. 1978. *Evolution of the earth's crust.* Academic Press.

Jordan, Thomas H. 1979. The deep structure of the continents. *Scientific American* 240 (January): 92–107.

Strangway, David W. 1980. *The Continental crust and its mineral deposits.* Special Paper No. 20, Geological Society of Canada.

Jones, David L., Allan Cox, Peter Coney and Myrl Beck. 1982. The growth of western North America. *Scientific American* 247 (November): 70–84.

Windley, Brian F. 1984. *The evolving continents,* 2nd ed. John Wiley & Sons, Inc.

Hoffman, P. F. 1988. United plates of America, the birth of a craton: Early Proterozoic assembly and growth of Laurentia. *Annual Review of Earth and Planetary Sciences* 16:543–603.

Bally, A. W., and A. R. Palmer, eds. 1989. Overview. In *The geology of North America,* vol. A. Geological Society of America.

2. The Oceanic Crust

Heezen, Bruce C., and Charles D. Hollister. 1971. *The face of the deep.* Oxford University Press.

Le Pichon, Xavier, and Jean Francheteau. 1973. *Plate tectonics.* Elsevier-Scientific Publishing Company.

Ballard, Robert D., and James G. Moore. 1977. *Photographic atlas of the Mid-Atlantic Ridge rift valley.* Springer-Verlag.

Sclater, J. G., and C. Tapscott. 1979. The history of the Atlantic. *Scientific American* 240 (June): 156–174.

Francheteau, Jean, Thierry Juteau, David Needham and Claude Rangin. 1980. *Birth of an ocean: The crest of the East Pacific Rise.* Centre National pour l'Exploitation des Océans.

Macdonald, K. C., and B. P. Luyendyk. 1981. The crest of the East Pacific Rise. *Scientific American* 244 (May): 100–116.

Menard, H. W. 1986. *The ocean of truth: A personal history of global tectonics.* Princeton University Press.

Vogt, Peter R., and B. E. Tucholke, eds. The western North Atlantic region. In *The geology of North America,* vol. M. Geological Society of America.

Winterer, E. L., D. M. Hussong and R. W. Decker, eds. 1989. The eastern Pacific Ocean and Hawaii. In *The geology of North America,* vol. N. Geological Society of America.

3. The Earth's Hot Spots

Smith, R. B., and R. L. Christiansen. 1980. Yellowstone Park as a window on the earth's interior. *Scientific American* 242 (February): 104–116.

Vogt, Peter R. 1983. The Iceland mantle plume: Status of the hypothesis after a decade of new work. In *Structure and Development of the Greenland-Scotland Ridge,* eds. Martin H. Bott and Svend Saxov. Plenum Publishing Corp.

Morgan, W. Jason. 1983. Hotspot tracks and the early rifting of the Atlantic. *Tectonophysics* 94 (May 1): 123–139.

Vink, Gregory E. 1984. A hotspot model for Iceland and the Vøring Plateau. *Journal of Geophysical Research* 89 (November 10): 9949–9959.

Engebretson, D. C., A. Cox and R. G. Gordon. 1985. *Relative motions between oceanic and continental plates in the Pacific Basin.* Special Paper 206, Geological Society of America.

4. Seismic Tomography

Nakanishi, Ichiro, and Don L. Anderson. 1982. Worldwide distribution of group velocity of mantle Rayleigh waves as determined by spherical harmonic inversion. *Bulletin of the Seismological Society of America* 72 (August): 1185–1194.

McKenzie, D. P. 1983. The earth's mantle. *Scientific American* 249 (September): 66–78.

Dziewonski, Adam M. 1984. Mapping the lower mantle: Determination of lateral heterogeneity in P velocity up to degree and order 6. *Journal of Geophysical Research* 89 (July 10): 5929–5952.

Nataf, H.-C., and Don L. Anderson. 1984. Anisotropy and shear-velocity heterogeneities in the upper mantle. *Geophysical Research Letters* 11 (February): 109–112.

Woodhouse, John H., and Adam M. Dziewonski. 1984. Mapping the upper mantle: Three-dimensional modelling of earth structure by inversion of seismic waveforms. *Journal of Geophysical Research* 89 (July 10): 5953–5986.

Fischer, K. M., T. H. Jordan and K. C. Creager. 1988. Seismic constraints on the morphology of deep slabs. *Journal of Geophysical Research* 93:4773–4783.

5. How Continents Break Up

Courtillot, V., A. Galdeano and J. L. Le Mouël. 1980. Propagation of an accreting plate boundary: A discussion of new aeromagnetic data in the Gulf of Tadjurah and southern Afar. *Earth and Planetary Science Letters* 47 (March): 144–160.

Hey, Richard, Frederick K. Duennebier and W. Jason Morgan. 1980. Propagating rifts on midocean ridges. *Journal of Geophysical Research* 85 (July 10): 3647–3658.

Atwater, Tanya. 1981. Propagating rifts in seafloor spreading patterns. *Nature* 290 (March 19): 185–186.

Hey, R. N., and D. S. Wilson. 1982. Propagating rift explanation for the tectonic evolution of the northeast Pacific: The pseudomovie. *Earth and Planetary Science Letters* 58 (April): 167–188.

Courtillot, V. 1982. Propagating rifts and continental breakup. *Tectonics* 1 (June): 239–250.

Vink, G. E., W. J. Morgan and W.-L. Zhao. 1984. Preferential rifting of continents: A source of displaced terranes. *Journal of Geophysical Research* 89: 10072–10077.

Bonatti, E. 1987. The rifting of continents. *Scientific American* 256 (March): 96–103.

White, R. S., and D. P. McKenzie. 1989. Volcanism at rifts. *Scientific American* 261 (July): 62–71.

6. Oceanic Fracture Zones

Wilson, J. Tuzo. 1965. A new class of faults and their bearing on continental drift. *Nature* 207 (July 24): 343–347.

Sleep, Norman H., and Shawn Biehler. 1970. Topography and tectonics at the intersections of fracture zones with central rifts. *Journal of Geophysical Research* 75 (May 10): 2748–2752.

Bonatti, Enrico. 1978. Vertical tectonism in oceanic fracture zones. *Earth and Planetary Science Letters* 37 (January): 369–379.

Fujita, Kazuya, and Norman H. Sleep. 1978. Membrane stresses near mid-ocean ridge-transform intersections. *Tectonophysics* 50 (September 20): 207–221.

Bonatti, Enrico, and Andy Chermak. 1981. Formerly emerging crustal blocks in the equatorial Atlantic. *Tectonophysics* 72 (February 10): 165–180.

Fox, P. J., and D. G. Gallo. 1984. A tectonic model for ridge-transform-ridge plate boundaries: Implications for the structure of oceanic lithosphere. *Tectonophysics* 104:205–242.

7. Terranes

Silbering, N. J., and J. Hillhouse. 1977. Wrangellia: A displaced terrane in northwestern North America. *Canadian Journal of Earth Sciences* 14:2565–2577.

Coney, Peter J., David L. Jones and James W. H. Monger. 1980. Cordilleran suspect terranes. *Nature* 288 (November 27): 329–333.

Williams, Harold, and Robert D. Hatcher, Jr. 1982. Suspect terranes and accretionary history of the Appalachian orogen. *Geology* 10 (October): 530–536.

Jones, David L., Allan Cox, Peter Coney and Myrl Beck. 1982. The growth of western North America. *Scientific American* 247 (November): 70–84.

Reymer, A., and G. Schubert. 1984. Phanerozoic addition rates to the continental crust and crustal growth. *Tectonics* 3 (February): 63–77.

Howell, D. G., ed. 1985. *Tectonostratigraphic terranes of the circumpacific region.* Earth Science Series No. 1, Circum-Pacific Council for Energy and Mineral Resources.

Hillhouse, J. W., ed. 1989. *Deep structure and past kinematics of accreted terranes.* Geophysical Monograph 50, American Geophysical Union.

8. Ophiolites

Coleman, Robert G. 1977. *Ophiollites: Ancient oceanic lithosphere?* Springer-Verlag.

Panayiotou, A., ed. 1980. *Ophiolites: Proceedings of an international ophiolite symposium held in Nicosia, Cyprus, in 1979.* Geological Survey Department, Nicosia, Cyprus.

Emiliani, C., ed. 1981. The oceanic lithosphere. In *The Sea*, vol. 7. John Wiley & Sons, Inc.

Moores, E. M. 1982. Origin and emplacement of ophiolites. *Reviews of Geophysics and Space Physics* 20:735–760.

Varga, R. J., and E. M. Moores. 1985. Spreading structure of the Troodos ophiolite, Cyprus. *Geology* 13:846–850.

Gass, I. G., S. J. Lippard and A. W. Shelton. 1986. *Ophiolites and oceanic lithosphere.* Geological Society Special Publication No. 13. Blackwell's.

9. The Structure of Mountain Ranges

Molnar, Peter, and Paul Tapponnier. 1978. Active tectonics of Tibet. *Journal of Geophysical Research* 83 (November 10): 5361–5375.

Lyon-Caen, H., and Peter Molnar. 1983. Constraints on the structure of the Himalaya from an analysis of gravity anomalies and a flexural model of the lithosphere. *Journal of Geophysical Research* 88 (October 10): 8171–8191.

Karner, G. D., and A. B. Watts. 1983. Gravity anomalies and flexure of the lithosphere at mountain ranges. *Journal of Geophysical Research* 88 (December 10): 10449–10477.

Suárez, Gerardo, Peter Molnar and B. C. Burchfiel. 1983. Seismicity, fault plane solutions, depth of faulting, and active tectonics of the Andes of Peru, Ecuador, and southern Colombia. *Journal of Geophysical Research* 88 (December 10): 10403–10428.

Clark, S. P., B. C. Burchfiel and J. Suppe, eds. 1988. *Processes in continental lithospheric deformation.* Special Paper 218, Geological Society of America.

10. The Southern Appalachians and the Growth of Continents

Sougy, J. 1962. West African fold belt. *Geological Society of America Bulletin* 73 (July): 871–876.

Bird, John M., and John F. Dewey. 1970. Lithosphere plate: Continental margin tectonics and the evolution of the Appalachian orogen. *Geological Society of America Bulletin* 81 (April): 1031–1060.

Hatcher, Robert D., Jr. 1978. Tectonics of the western Piedmont and Blue Ridge, southern Appalachians: Review and speculation. *American Journal of Science* 278 (March): 276–304.

Cook, Frederick A., Dennis S. Albaugh, Larry D. Brown, Sidney Kaufman, Jack E. Oliver and R. D. Hatcher, Jr. 1979. Thin-skinned tectonics in the crystalline southern Appalachians: COCORP seismic-reflection profiling of the Blue Ridge and Piedmont. *Geology* 7 (December): 563–567.

Williams, H. 1984. Miogeoclines and suspect terranes of the Caledonian-Appalachian orogen: Tectonic patterns in the North Atlantic region. *Canadian Journal of Earth Sciences* 21:887–901.

Hatcher, R. D., Jr. 1987. Tectonics of the southern and central Appalachian internides. *Annual Reviews of Earth Planetary Sciences* 15:337–362.

11. The Growth of Western North America

McElhinny, M. W. 1973. *Paleomagnetism and plate tectonics.* Cambridge University Press.

Jones, D. L., N. J. Silbering, Béla Csejtey, Jr., W. H. Nelson and Charles D. Blome. 1980. *Age and structural significance of ophiolite and adjoining rock in the Upper Chulitna District, south-central Alaska.* U.S. Geological Survey Professional Paper 1121-A. U.S. Government Printing Office.

Coney, Peter J., David L. Jones and James W. H. Monger. 1980. Cordilleran suspect terranes. *Nature* 288 (November 27): 329–333.

Beck, Myrl E., Jr. 1980. Paleomagnetic record of plate-margin tectonic processes along the western edge of North America. *Journal of Geophysical Research* 85 (December 10): 7115–7131.

Ben-Avraham, Z., A. Nur, D. Jones and A. Cox. 1981. Continental accretion: From oceanic plateaus to allochthonous terranes. *Science* 213 (July 3): 47–54.

Monger, J. W. H. 1984. Cordilleran tectonics: A Canadian perspective. *Bulletin de la Societe geologique de France* 26:255–278.

Allmendinger, R. W., T. A. Hauge, E. C. Hauser, C. J. Potter, S. L. Klemperer, K. D. Nelson, P. Knuepfer and J. Oliver. 1987. Overview of the COCORP 40° N transect, western U.S.A. *Geological Society of America Bulletin* 98:308–319.

Debiche, M. G., A. Cox and D. Engebretson. 1987. *The motion of allochthonous terranes across the Pacific Basin.* Special Paper 207, Geological Society of America.

12. The Supercontinent Cycle

Fisher, Alfred G. 1983. The two Phanerozoic supercycles. In *Catastrophes and Earth History*, eds., W. A. Berggren and John Van Covering. Princeton University Press.

Worsley, Thomas R., Damian Nance and Judith B. Moody. 1984. Global tectonics and eustasy for the past 2 billion years. *Marine Geology* 58 (July 25): 373–400.

Nance, R. Damian, Thomas R. Worsley and Judith B. Moody. 1986. Post-archean biochemical cycles and long-term episodicity in tectonic processes. *Geology* 14 (June): 514–518.

Worsley, Thomas R., R. Damian Nance and Judith B. Moody. 1986. Tectonic cycles and the history of the earth's biochemical and paleoceanographic record. *Paleoceanography* 1 (September): 233–263.

Hoffman, P. F. 1989. Speculations on Laurentia's first gigayear (2.0 to 1.0 Ga). *Geology* 14:135–138.

Sources of the Photographs

EROS Data Center: Figure 1.1
B. Clark Burchfield: Figure 1.7

Jean Francheteau: Figures 2.2 and 2.3
William F. Haxby, Lamont-Doherty Geological Observatory: Figures 2.4 and 2.6

EROS Data Center: Figure 5.3
Vincent Courtillot, University of Paris: Figure 5.6

William F. Haxby, Lamont-Doherty Geological Observatory: Figure 6.4
Bonatii, Enrico, Lamont-Doherty Geological Observatory: Figure 6.6
Dee Breger, Lamont-Doherty Geological Observatory: Figure 6.7

EROS Data Center: Figure 7.1
Benita L. Murchey and David L. Jones, U.S. Geological Survey: Figure 7.2
Anita G. Harris, U.S. Geological Survey: Figure 7.3

Earth Satellite Corporation: Figure 8.1
Ian G. Gass, Open University: Figures 8.3 and 8.4

National Aeronautics and Space Administration: Figure 9.1
Peter Molnar: Figure 9.7
Servicio Aerofotgráfico Nacional del Peru: Figure 9.8

National Aeronautics and Space Administration: Figure 10.1

David L. Jones, U.S. Geological Survey: Figure 11.6
Andrew Tomko: Figure 11.7

INDEX

Page numbers in *italics* indicate illustrations.